Millimetre Wave Antennas for Gigabit Wireless Communications

Millimetre Wave Antennas for Gigabit Wireless Communications

A Practical Guide to Design and Analysis in a System Context

Kao-Cheng Huang

University of Greenwich, UK

David J. Edwards

University of Oxford, UK

A John Wiley and Sons, Ltd, Publication

Library of Congress Cataloging-in-Publication Data

Huang, Kao-Cheng.
  Millimetre wave antennas for gigabit wireless communications : a practical guide
  to design and analysis in a system context / Kao-Cheng Huang, David J. Edwards.
    p. cm.
  Includes bibliographical references and index.
  ISBN 978-0-470-51598-3 (cloth)
  1. Microwave antennas. 2. Gigabit communications. 3. Milimeter waves.
  I. Edwards, David J. II. Title.
  TK7871.67.M53H83 2008
  621.382′4—dc22
                                                        2008013165

A catalogue record for this book is available from the British Library

ISBN 978-0-470-51598-3 (HB)

Set in 10/12pt Times by Integra Software Services Pvt. Ltd, Pondicherry, India
Printed in Singapore by Markono Print Media Pte Ltd

Contents

Preface

This book presents antenna design and analysis at the level to produce an understanding of the interaction between a wireless system and its antenna, so that the overall performance can be predicted. Gigabit wireless communications require a considerable amount of bandwidth, which can be supported by millimetre waves. Millimetre wave technology has now come of age, and at the time of writing the standards of IEEE 802.15.3c, *WirelessHD*™ and *ECMA* are on schedule to be finalised. The technology has attracted new commercial wireless applications and new markets, such as the capacity for high-speed downloading and wireless high-definition TVs. This book summarises and reports the extensive research over recent years and emphasises the importance and requirements of antennas for gigabit wireless communications, with an emphasis on wireless communications in the 60 GHz ISM band and in the E-band. This book I reviews the particular requirements for this application and addresses the design and feasibility of millimetre wave antennas; such as planar antennas, rod antennas and antenna arrays. Examples of designs are included, along with a detailed analysis of their performance. In addition, this book includes a bibliography of current research literature and patents in this subject area. Finally, the applications of these antennas are discussed in the light of different forthcoming wireless standards.

Millimetre Wave Antennas for Gigabit Wireless Communications endeavours to offer a comprehensive treatment of antennas based on electronic consumer applications, providing a link to applications of computer-aided design tools and advanced materials and technologies. The major features of this book include a discussion of the many novel millimetre wave antenna configurations available with newly reported design techniques and methods.

Although it contains some introductory material, this book is intended to provide a collection of millimetre wave antenna design considerations for communication system designers and antenna designers. The book should also act as a reference for postgraduate students, researchers and engineers in millimetre wave engineering and an introduction to the various design considerations. It can also be used for millimetre wave teaching. A summary of each chapter is given below.

Chapter 1 introduces the near-term developments in millimetre wave communications. The importance and requirements of millimetre wave antennas are discussed based on channel performance, link budget, and applications in line-of-sight and non-line-of-sight scenarios. Sections addressing system-level considerations include references to subsequent chapters that contain a more detailed treatment of antenna design.

Chapters 2 to 8 address conventional configurations of millimetre wave antennas.

Chapter 2 considers several critical factors that limit the performance of millimetre wave antennas. As the antenna design has become critical in wireless communications, the limitations of antenna design are also discussed in this chapter.

Chapter 3 describes the variety of millimetre wave planar antennas, and lists basic feeding methods and useful references on a wide variety of techniques for producing low-profile antennas.

Chapter 4 deals with millimetre wave integrated horn antennas. The chapter includes a discussion of circular polarisation optimisation techniques, such as those for array antennas. With circular waveguide modes that can be used for mode tracking described in Chapter 8.

Chapter 5 addresses low cost and high directivity of millimetre wave rod antennas. Different feeding methods, maximum gain, and beam tilting are discussed in detail. With multiple-rod antennas that can be used as beam-switching antennas discussed in Chapter 7.

Chapter 6 describes the variety of millimetre wave lens antennas, relevant feeding methods and novel architectures. Lens antennas, with the advantages of light weight and small height, are identified as designs that can be used for new applications.

Chapter 7 discusses millimetre wave multibeam antennas and their construction. Novel antennas with advanced radiation characteristics have been demonstrated. Some of the effects of mutual coupling of signals and noises between array elements are covered. This interaction modifies the active array element patterns and can cause impedance changes during scanning.

Chapter 8 focuses on smart antennas and their usage in wireless communications. Wide-ranging technologies such as beam switching, beam steering, MIMO and mode tracking, that satisfy special needs are considered. These technologies could produce low-profile high-gain electronic scanning systems in conjunction with the antenna elements described in Chapters 3 to 7.

Chapter 9 explores millimetre wave antenna materials and manufacturing techniques. Materials technologies are discussed such as LTCC, LCP, CMOS, high-temperature super-conductors, carbon nanotubes, etc. New materials offer new design concepts and promise future exciting antenna technology trends.

Finally, chapter 10, extrapolates the wireless applications of millimetre wave antennas in a envisaged future market. This book only briefly addresses the details of electromagnetic analysis, with the fundamentals of the subject requiring a more detailed study than can be given in this system design-oriented book.

First of all, the authors wish to acknowledge the copyright permission from IEEE (US), European Microwave Association (Belgium), John Wiley & Sons, Inc. (US), ERA (UK) and Su Khiong Yong (US).

The authors are indebted to many researchers for their published works, which were rich sources of reference upon which this book reports and summarises. Their sincere gratitude extends to the Editor, Sarah Hinton, and the reviewers for their support in the writing of this book. The help provided by Tiina Ruonamaa and other members of the staff at John Wiley & Sons, Ltd is most appreciated. The authors also wish to thank their colleagues at the University of Oxford, and University of Greenwich.

In addition, Kao-Cheng Huang would like to thank Prof. Mook-Seng Leong, National University of Singapore (Singapore), Prof. Rüdiger Vahldieck, ETH (Switzerland), Prof. Ban-Leong Ooi, National University of Singapore (Singapore), Dr David Haigh, Imperial College (UK), Prof. Francis Lau, Hong Kong Polytechnic University (China), Dr H.M. Shen, University of Edinburgh (UK), Dr Chris Stevens, University of Oxford (UK) and Dr Jia-Sheng Hong, Heriot-Watt University (UK) for their many years of support. David Edwards would also like to thank Charlotte Edwards for her help in the final stages of the book. *Note*: During the later stages of the production of this book Dr Kao-Cheng Huang was taken seriously ill. The book has been completed from his notes and we apologise for any resulting omissions.

List of Abbreviations

A/V	audio/visual
ADC	analogue-to-digital conversion
AP	access points
AR	axial ratio
ARIB	Association of Radio Industries and Business
ASP	aperture stacked patch
BER	bit error rate
CB-FGC	conductor-backed finite ground coplanar
CBCPW	conductor-backed coplanar waveguide
CCS	complementary conducting strip
CEPT	European Conference of Postal and Telecommunications Administrations
CPS	coplanar stripline
CTE	coefficient of thermal expansion
DAS	distributed antenna systems
DBF	digital beamforming
DLA	discrete lens array
DoA	direction of arrival
DRA	dielectric resonator antenna
EBGs	electromagnetic bandgaps
ECC	Electronic Communications Committee
ECMA	European Computer Manufacture Association
EIRP	equivalent isotropic radiated power
ERC	European Radiocommunications Committee
ESPRIT	Estimation of Signal Parameters via Rotational Invariance Techniques
ETSI	European Telecommunications Standards Institute
FCC	Federal Communication Commissions
FDA	Food and Drug Administration
FDTD	finite-difference time-domain
FLA	filter–lens array
FPC	Fabry–Perot cavity
FT	Fourier transform
GO	geometric optics
GSM	Global System for Mobile Communications
HDMI	high-definition multimedia interface

HDTV	high-definition television
HEM	hybrid electromagnetic
HPBW	half-power beamwidth
HTS	high-temperature superconductors
IC-SMT	Industry Canada Spectrum Management and Telecommunications
ISM	industrial, scientific and medical
ISPs	Internet Service Providers
IVC	Inter-Vehicle Communications
LA	lens array
LCP	liquid crystal polymer
LHCP	left-hand circular polarisation
LHM	left-handed materials
LNA	low-noise amplifier
LOS	line-of-sight
LPD	low probability of detect
LPI	low probability of intercept
LTCC	low-temperature co-fired ceramic
MANETs	mobile ad hoc networks
MCM	multichip modules
MEMS	microelectromechanical system
MIMO	multi-input multi-output
MMIC	monolithic-microwave integrated circuit
MPHPT	Ministry of Public Management, Home Affairs, Posts, and Telecommunications
MSK	minimum shift keying
MT	mobile terminal
NLOS	non-line-of-sight
OFDM	orthogonal frequency division multiplexing
OMT	orthomode transducer
OOK	on/off keying
PA	1. power amplifier
	2. phased array
PAPR	peak-to-average power ratio
PCB	printed circuit board
PDA	personal data assistant
PHY	physical layer
PMP	portable media player
PRS	partially reflective surface
PS	portable station
PTHs	plated through holes
QPSK	quadrature phase-shift keying
RAUs	radio access units
RF	radio frequency
RHCP	right-hand circular polarisation
RRH	remote radio heads
SC	single carrier
SCBT	single-carrier block transmission

SIR	signal-to-interference
SNR	signal-to-noise ratio
SoC	system-on-chip
SoP	system-on-package
SP3T	single-pole triple-throw
SSFIP	strip slot foam inverted patch antenna
ULA	uniform linear arrays
UWB	ultra-wideband
VCC	voltage-controlled oscillator
WLANs	wireless local area networks
WMN	wireless mesh network
WPANs	wireless personal area networks

1

Gigabit Wireless Communications

The demand for high data rate and high integrity services seems set to grow for the foreseeable future. In this chapter the basic ideas and application areas for gigabit Ethernet are introduced, and the requirements for high-performance networks are described. The role of the antenna in these systems is addressed, and consideration of the performance parameters outlined.

This chapter is organised as follows. Section 1.1 describes a number of application scenarios and highlights the requirements for a specific application, namely uncompressed high-definition video streaming. Section 1.2 describes the worldwide regulatory efforts and standardisation activities. Section 1.3 presents the characteristics of millimetre waves. Section 1.4 presents measured propagation results and channel performance. Section 1.5 describes system design and performance. Section 1.6 discusses the role of the antenna within the system and the technical challenges that need to be resolved for the full deployment of 60 GHz radio networks. Section 1.7 describes the link budget, which is pivotal in determining the performance of the system. In this section noise is also examined, and its impact on link behaviour. Section 1.8 summarises the main points of the chapter.

1.1 Gigabit Wireless Communications

The adoption of each successive generation of Ethernet technology has been driven by economics, performance demand, and the rate at which the price of the new generation has approached that of the old. As the cost of 100 Mbps Ethernet decreased and approached the previous cost of 10 Mbps Ethernet, users rapidly moved to the higher performance standard. In January 2007, 10 gigabit Ethernet over copper wiring was announced by the industry [1]. Additionally, gigabit Ethernet became economic (e.g. below $200) for server connections, and desktop gigabit connections have come within $10 or less of the cost of 100 Mbps technology. Consequently, gigabit Ethernet has become the standard for servers, and systems are now routinely ordered with gigabit Network interface cards. Mirroring events in the wired world, as the prices of wireless gigabit links approach the prices of 100 Mbps links, users are switching to the higher-performance product, both for traditional wireless applications, as well as for applications that only become practical at gigabit speeds.

Millimetre Wave Antennas for Gigabit Wireless Communications Kao-Cheng Huang and David J. Edwards
© 2008 John Wiley & Sons, Ltd

In terms of a business model, wireless communications have pointed towards an approaching need for gigabit speeds and longer-range connectivity as the applications emerge for home audio/visual (A/V) networks, high-quality multimedia, voice and data services. Current wireless local area networks (WLANs) offer peak rates of 54 Mb/s, with 200–540 Mb/s, such as IEEE 802.11n, becoming available soon. However, even 500 Mb/s is inadequate when faced with the demand for higher access speed from rich media content and competition from 10 Gb/s wired LANs. In addition, future home A/V networks will require a Gb/s data rate to support multiple high-speed, high-definition A/V streams (e.g. carrying an uncompressed high-definition video at resolutions of up to 1920×1080 progressive scans, with latencies ranging from 5 to 15 ms) [2].

Based on the technical requirements of applications for high-speed wireless systems, both industry and the standardisation bodies need to take into account the following issues:

1. Pressure on data rate increases will persist.
2. There is a need for advanced domestic applications such as high-definition wireless multimedia, which demand higher data rates.
3. Data streaming and download/memory back-up times for mobile and personal devices will also place demands on the shared resource, and user models point to very short dwell times for these downloads.

Some approaches, such as IEEE 802.11n, are improving data rates by evolving the existing WLANs standards to increase the data rate; to up to 10 times faster than IEEE 802.11a or 802.11g. Others, such as the ultra-wideband (UWB) are pursuing much more aggressive strategies, such as sharing spectra with other users. Another approach that will no doubt be taken will be the time-honoured strategy of moving to higher, unused and unregulated millimetre wave frequencies.

Despite millimetre wave technology having been established for many decades, the millimetre wave systems available have mainly been deployed for military applications. With the advances in process technologies and low-cost integration solutions, this technology has started to gain a great deal of momentum from academia, industry and standardisation bodies. In very broad terms, millimetre wave technology can be classified as occupying the electromagnetic spectrum that spans between 30 and 300 GHz, which corresponds to wavelengths from 10 to 1 mm. In this book, the main focus will be on the 60 GHz industrial, scientific and medical (ISM) band (unless otherwise specified, the terms "60 GHz" and "millimetre wave" will be used interchangeably), which has emerged as one of the most promising candidates for multigigabit wireless indoor communication systems.

Although the IEEE 802.11n standard will improve the robustness of wireless communications, only a modest increase in wireless bandwidth is provided and the data rate is still lower than 1 Gb/s. Importantly, 60 GHz technology offers various advantages over currently proposed or existing communications systems. One of the deciding factors that makes 60 GHz technology attractive and has prompted significant interest recently, is the establishment of (relatively) huge unlicensed bandwidths (up to 7 GHz) that are available worldwide. The spectrum allocations are mainly regulated by the International Telecommunication Union. The details for band allocation around the world can be found in Section 1.2.

While this is comparable to the unlicensed bandwidth allocated for ultra-wideband purposes (~2–10 GHz), the 60 GHz band is continuous and less restricted in terms of power limits (also there are less existing users). This is due to the fact that the UWB system is an overlay system and thus subject to different considerations and very strict regulation. The large band at 60 GHz is in fact one of the largest unlicensed spectral resources allocated in history. This huge bandwidth offers potential in terms of capacity and flexibility and makes 60 GHz technology particularly attractive for gigabit wireless applications. Although 60 GHz regulations allow much higher transmit power compared to other existing wireless local area networks (e.g. maximum 100 mW for IEEE 802.11 a/b/g) and wireless personal area network (WPAN) systems, the higher transmit power is necessary to overcome the higher path loss at 60 GHz (see Table 1.1).

Table 1.1 Path loss and transmit power comparison for different wireless standards

	10 m path loss (dB)	Maximum transmit power (mW)
802.11a	66	40
802.11b/g	60	100
802.15.3c	88	500

In addition, the typical 480 Mbps bandwidth of UWB cannot fully support broadcast video and therefore the data packets need to be recompressed. This forces manufacturers to utilise expensive encoders and more memory into their systems, in effect losing video content and adding latency in the process. Therefore, 60 GHz technology could actually provide better resolution, with less latency and cost for television, DVD players and other high-definition equipment, compared to UWB.

Taking into consideration the development of consumer electronics, currently the IEEE 802.15.3c standard [3] provides 1–3 Gb/s wireless personal area network solutions, projected for introduction in the years 2008 to 2009. Also, WiMedia 2.0 [4], which can be used for large file transfer applications, is to be developed, so the target is to have a data rate of 5 Gb/s or higher raw bit rates and with more than a 10 m range for indoor applications.

Figure 1.1 shows the development and the trend of wireless standards. Advanced wireless technology should always adopt timelines/milestones to increase data rates by ∼5 to 10 times every 3 to 4 years to keep up with the pace of projected demand.

While the high path loss seems to be a disadvantage at 60 GHz, it does however confine the 60 GHz power and system operation in an indoor environment. Hence, the effective interference levels for 60 GHz are less severe than those systems located in the congested 2–2.5 GHz and 5–5.8 GHz regions. In addition, higher frequency re-use can also be achieved over a very short distance in an indoor environment, thus allowing a very high throughput network. The compact size of the 60 GHz radio also permits multiple antenna solutions at the user terminal that are otherwise difficult, if not impossible, at lower frequencies. Compared to a 5 GHz system, the form factor of millimetre wave systems is approximately 140 times smaller and can be conveniently integrated into consumer electronic products, but it will require new design methodologies to meet modern communication needs.

Designing a very high-speed wireless link that offers good quality-of-service and range capability presents a significant research and engineering challenge. Ignoring fading for the moment, in theory, the 1 Gb/s data rate requirement can be met, if the product of bandwidth

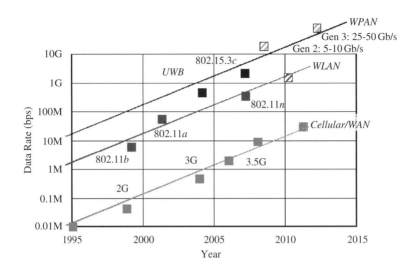

Figure 1.1   Data rate projections over time [5]

(in units of Hz) and spectral efficiency (in b/s per Hz units) equals 10^9. As shall be described in the following sections, a variety of cost, technology and regulatory constraints make such a solution very challenging.

Despite the various advantages offered, millimetre wave based communications suffer a number of critical problems that must be resolved. Figure 1.2 shows the data rates and range requirements for a number of WLAN and WPAN systems. Since there is a need to distinguish between different standards for broader market exploitation, the IEEE 802.15.3c standard is positioned to provide gigabit rates and a longer operating range. At these rates and ranges, it will be a difficult task for millimetre wave systems to provide a sufficient power margin to ensure a reliable communication link. Furthermore, the delay spread of the channel under consideration is another limiting factor for high-speed transmission. Large delay spread values can easily increase the complexity of the system beyond the practical limit for equalisation [6].

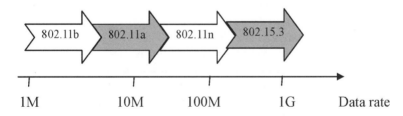

Figure 1.2   Data rates and range requirements for WLAN and WPAN standards and applications. Millimetre wave technology, i.e. IEEE 802.15.3c, is aiming for very high data rates [6]

If a 10 mW power input to the antenna is assumed with a 10 dBi gain based on a highly integrated, low-cost design with a steerable beam at 60 GHz, a Shannon capacity curve is produced, as shown in Figure 1.3. The formula used to derive these curves is presented in Equations (1.3) and (1.4) in Section 1.4.

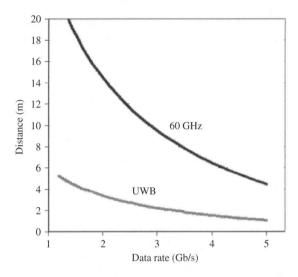

Figure 1.3 Shannon's capacity curve in a 1 GHz occupied bandwidth for 60 GHz versus UWB (noise figure is set at 8 dB) [5]

In the search for the provision of higher data rates, radio systems have tended to look at higher frequencies where an unregulated spectrum is available. As an alternative, a (free space) wireless optical LAN also competes as one of the communication technologies that are able to offer a significant unregulated spectrum. Diffuse optical networks use wide- angle sources and scatter from surfaces in the room to provide optical 'ether' similar to that which would be obtained using a local radio transmitter [7]. This produces coverage that is robust to blocking, but the multiple paths between the source and receiver cause dispersion of the channel, thus limiting its performance. Additionally optical transmitters launch extremely high power, and dynamic equalisation is required for high bandwidth operation.

Optical networks have the potential to offer significant advantages over radio approaches, within buildings or in spaces with limited coverage. Many current systems use directed line-of-sight paths between transmitter and receiver [8]. These can provide data rates of hundreds of megabits per second and above, depending on particular parameters. However, the coverage area provided by a single channel can be quite small, so that providing area coverage, and the ability to roam, presents a major challenge. Line-of-sight channels can also be blocked, as there is no alternative scattered path between the transmitter and receiver, and this presents a major challenge in network design [7]. Multiple-base stations within a room would provide coverage in this case, and optical or fixed connections could be used between the stations. A commercial line-of-sight system is currently offered by Victor Company of Japan, Limited (JVC), giving 10 Mb/s Ethernet connections [9].

In general, optical channels are subject to eye safety regulation, which is difficult to meet, particularly for line-of-sight channels [7]. Typically optical LANS work in the near- infrared region (between 700 and 1000 nm) where optical sources and detectors are low cost and regulations are particularly strict. At longer wavelengths (1500 nm and above) the regulations are much less stringent, although sources at this wavelength and power output are not widely available [10].

As previously mentioned, the other major problem for optical channels is that of blocking. Line-of-sight channels in particular are required for high-speed operation and these are by their nature subject to blocking. Within a building, networks must be designed using appropriate geometries to avoid blocking, and this is usually solved by using multiple access points to allow complete coverage [10, 11].

Table 1.2 compares the characteristics of three technologies for gigabit communications: UWB radio, millimetre wave and wireless optics.

Table 1.2 Comparison of three new technologies for gigabit wireless communications [12–14]

	Millimetre wave	UWB radio	Optical wireless
Advantage	1. High data rates (up to Gb/s) 2. Compatible with fibre optic networks at 60 GHz	1. Low power 2. Short range 3. Low data 4. Penetration through obstacles in the transmission path	1. High data rate 2. Unlicensed and unregulated.
Challenge	1. Low cost 2. Low power	1. Matched filter problem 2. Antenna parameter trade-off	1. Atmospheric loss ranging from 10 dB/km(sunny) to 350 dB/km(foggy) 2. Multi-user application 3. No protection for the link
Peer-to-peer	Indoor/outdoor	Indoor/outdoor	Indoor/outdoor
Multiple-access	Indoor/outdoor	Indoor	Indoor
Data rate	>1.25 Gb/s at 60 GHz ~10 Gb/s at 122.5 GHz	500 Mbps within 10 m (FCC)	~1.25 Gb/s (peer-to-peer)
Indoor maximum range	Room area	76 m (station in commercial building)	7 m (mobile) 10 m (station)
DC power consumption	High	Low	DC 5 V, 500 mA (mobile)
Maximum TX power	500 mW (FCC 15.255)	Maximum output power of 1 W spread over spectrum Maximum power density: −41.3 dBm/MHz (FCC)	Power density should be less than 1 mW/cm^2 (FDA)
Notes	Antenna design is one of the main challenges	1. Infrastructure or peer-to-peer for indoor application 2. Only peer-to-peer for hand-held application (FCC)	Eye safety should be considered

1.2 Regulatory Issues

1.2.1 Europe

The European Telecommunications Standards Institute (ETSI) and European Conference of Postal and Telecommunications Administrations (CEPT) have been working closely to establish a legal framework for the deployment of unlicensed 60 GHz devices [15]. In general, the 59–66 GHz band has been allocated for mobile services without specific decision on the regulations, as shown in Figure 1.4. The CEPT Recommendation T/R 22–03 has provisionally recommended the use of the 54.25–66 GHz band for terrestrial and fixed mobile systems [16]. However, this provisional allocation has been recently withdrawn [6].

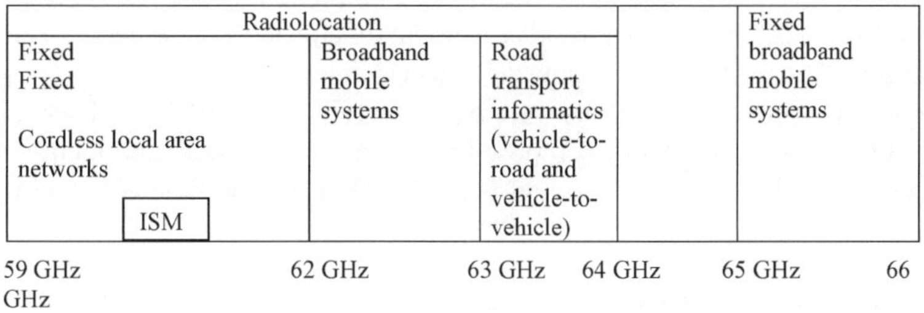

Radiolocation
Fixed
Fixed
Cordless local area networks
ISM
Broadband mobile systems
Road transport informatics (vehicle-to-road and vehicle-to-vehicle)
Fixed broadband mobile systems
59 GHz
62 GHz
63 GHz
64 GHz
65 GHz
66 GHz

Figure 1.4 The 60 GHz frequency spectrum in Europe (ISM: industry, science and medicine) [17]

In 2003, the European Radiocommunications Committee (ERC) within the European Conference of Postal and Telecommunications Administrations revised the European Table of Frequency Allocations and Utilisations [17]. The ERC also considered the use of the 57–59 GHz band for fixed services without requiring frequency planning [18]. Later, the Electronic Communications Committee (ECC) within the CEPT recommended the use of point-to-point fixed services in the 64–66 GHz band [19]. In the most recent development, the ETSI proposed 60 GHz regulations to be considered by the Electronic Communications Committee of the European Conference of Post and Telecommunications Administrations for WPAN applications [20]. Under this proposal, 9 GHz of unlicensed spectrum has been allocated for 60 GHz operation. This band represents the union of the bands currently approved and under consideration in the first quarter of 2007.

The frequency band being considered is 57–66 GHz. The spectrum allocation is shown in Figure 1.5 and Table 1.3. This is the amalgamation of the bands currently approved for

Europe (57–66 GHz)
Japan (59–66 GHz)
Australia (59.4–62.9 GHz)
USA (57–64 GHz)

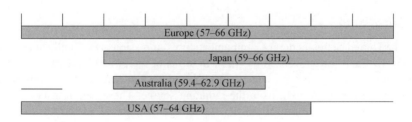

Figure 1.5 Geographically available 60 GHz spectrum and power

Table 1.3   International frequency allocation at 60 GHz [25]

Region	Unlicensed bandwidth (GHz)	Tx power	Maximum antenna gain	Reference
Europe	9 GHz (57–66) min 500 MHz	20 mW (max)	37 dBi	[20]
Japan	7 GHz (59–66) max 2.5 GHz	10 mW(max)	47 dBi	[22]
Korea	7 GHz (57–64)	10 mW(max)	To be decided	[23]
Germany	1 GHz (57.1–57.8) (58.6–58.9)	50 mW (max)	Not specified	[21]
USA	7 GHz (57–64)	500 mW (max)	Not specified	[24]

license-exempt use in Japan and the United States, and under proposed allocation in the Republic of China and the Republic of Korea. The existing etiquette rules, spectrum sharing studies and other analyses in these countries could be a model for considering the needs of commercial, military and scientific uses of these frequencies worldwide.

The proposed European Regulations were based on ETSI DTR/ERM-RM-049 [20]. It was proposed that the ECC considers the proposed regulation in Clause 6, and identifies the final frequency band for 60 GHz license-exempt operation. The proposed power level is shown in Table 1.4.

Table 1.4   Proposed power regulation [20, 26]

Minimum bandwidth	Maximum transmit power	Channel spacing	Notes
A minimum spectrum of 500 MHz is requested for the transmitted signal, which should, in theory and under the right circumstances, be able to share a spectrum with other users	+57 dBm EIRP (+20 dBm nominal with up to +37 dBi antenna gain or +10 dBm nominal with up to +47 dBi antenna gain)	No restriction	The transmit power is necessary to offset oxygen and material attenuation at this band, and is typical for gigabit commercial products in this band

In Germany, the regulatory requirements are that the frequency band of 57.1–57.8 and 58.6–58.9 GHz are used for a time-domain duplex (TDD) point-to-point connection. Its maximum EIRP (equivalent isotropic radiated power) is 15 dBW. The frequency band of 61–61.5 GHz is for location service and general use. The maximum EIRP is 10 W for the location service and 100 mW for general use [21].

1.2.2   United States

In 2001, the United States Federal Communication Commissions (FCC) allocated 7 GHz in the 54–66 GHz band for unlicensed use [24]. In terms of the power limits, FCC rules allow emission with an average power density of 9μ W/cm^2 at 3 m and maximum power density of 18μ W/cm^2 at a range of 3 m from the radiating source. These data translate to average equivalent isotropic radiated power (EIRP) and maximum EIRP of 40 and 43 dBm, respectively. The FCC also specified the total maximum transmit power of 500 mW for an emission bandwidth greater than 100 MHz. The devices must also comply with the radio frequency

(RF) radiation exposure requirements specified in Reference [24], Sections 1.307(b), 2.1091 and 2.1093. After taking the RF safety issues into account, the maximum transmit power is limited to 10 dBm. Furthermore, each transmitter must transmit at least one transmitter identification signal within a 1 s interval of the signal transmission. It is important to note that the 60 GHz regulations in Canada, which is regulated by Industry Canada Spectrum Management and Telecommunications (IC-SMT) [27], are harmonized with the US.

In October 2003, the FCC announced that the frequency bands from 71 to 76 GHz, 81 to 86 GHz and 92 to 95 GHz were available for wireless applications [28]. The FCC chairman heralded the ruling as opening a "new frontier" in commercial services and products [29]. The allocation provides the opportunity for a broad range of new products and services, including high-speed, point-to-point wireless local area networks and broadband Internet access at gigabit data rates and beyond.

The 70, 80 and 90 GHz allocations are significant. Collectively referred to as E-band, these three allocations are the highest frequencies ever licensed by the FCC. The nearly 13 GHz of allocated spectrum represents more bandwidth than all other previously existing commercial wireless spectrum combined. The ruling also permitted a novel licensing scheme, allowing cheap and fast frequency allocations to prospective users. All this was achieved at an unprecedented speed, from the initial petition to the formal release of the rules in scarcely more than two years.

1.2.3 Japan

In the year 2000, the Ministry of Public Management, Home Affairs, Posts, and Telecommunications (MPHPT) of Japan issued 60 GHz radio regulations for unlicensed utilization in the 59–66 GHz band [22]. The 54.25–59 GHz band is, however, allocated for licensed use. The maximum transmit power for the unlicensed use is limited to 10 dBm, with a maximum allowable antenna gain of 47 dBi. Unlike the arrangements in North America, the Japanese regulations specified that the maximum transmission bandwidth must not exceed 2.5 GHz. There is no specification for RF radiation exposure and transmitter identification requirements [22].

1.2.4 Industrial Standardisation

The first international industry standard that covered the 60 GHz band was the IEEE 802.16 standard for local and metropolitan area networks [30]. However, this is a licensed band and is used for line-of-sight (LOS) outdoor communications for last mile connectivity. In Japan, two standards related to the 60 GHz band were issued by the Association of Radio Industries and Business (ARIB), i.e. the ARIB-STD T69 and ARIB-STD T74 [31, 32]. The former is the standard for millimetre wave video transmission equipment for a specified low-power radio station (point-to-point system), while the latter is the standard for a millimetre wave ultra-high-speed WLAN for specified low-power radio stations (point-to-multipoint). Both standards cover the 59–66 GHz band defined in Japan (see Table 1.5).

Interest in the 60 GHz radio continued to grow with the formation of a Millimetre Wave Interest Group and Study Group within the IEEE 802.15 Working Group for WPAN. In March 2005, the IEEE 802.15.3c Task Group (TG3c) was formed to develop a millimetre wave-based alternative physical layer (PHY) for the existing IEEE 802.15.3 WPAN Standard 802.15.3-2003 [33]. The developed PHY is aimed to support a minimum data rate of 2 Gb/s over a few

Table 1.5 The 60 GHz standards in Japan

Code	Standard name	Note
ARIB STD-T69 (July 2004)	Millimetre-Wave Video Transmission Equipment for Specified Low Power Radio Station	Bandwidth: 1208 MHz Tx power: 10 dBm Rx antenna gain: 0 dBi
ARIB STD-T69 Revision (November 2005)	Millimetre-Wave Video Transmission Equipment for Specified Low Power Radio Station (only the part of the revision from Version 2.0 to 2.1)	
ARIB STD-T74 (May 2001)	Millimetre-Wave Data Transmission Equipment for Specified Low Power Radio Station (Ultra High Speed Wireless LAN System)	Bandwidth: 200 MHz Tx power: 10 dBm Rx antenna gain: 0 dBi
ARIB STD-T74 Revision (November 2005)	Millimetre-Wave Data Transmission Equipment for Specified Low Power Radio Station (Ultra High Speed Wireless LAN System) (only the part of the revision from Version 1.0 to 1.1)	

metres with optional data rates in excess of 3 Gb/s. This is the first standard that addresses multigigabit wireless systems and will form the key solution to many data rates serving applications, especially those related to wireless multimedia distribution. In other developments, WiMedia Alliance has recently announced the formation of the WiMedia 60 GHz Study Group with the aim of providing recommendations to the WiMedia Board of Directors on the feasibility issues related to 60 GHz technology. A decision will be taken in the near future about WiMedia's direction and involvement in the 60 GHz market.

In 2007, another group, WirelessHD™ (high definition), also released a specification that uses the unlicensed 60 GHz radio to send uncompressed HD video and audio at 5 Gb/s over distances of up to 30 feet, or within one room of a house. Its core technology promotes theoretical data rates up to 20 Gb/s, permitting it to scale to higher resolutions, colour depths and ranges. Coexisting with other wireless services, the Wireless HD platform is designed to operate cooperatively with existing, wireline display technologies. The specification maintains high-quality video, ensures the interoperability of consumer electronics devices, protects from signal interference and uses existing content protection techniques. The WirelessHD™ Group predicts that 60 GHz will allow the fast transmission speeds required for high-definition content.

In addition, the European Computer Manufacturers Association (ECMA International) Technical Committee Task Group (TG20) has also developed a standard for a 60 GHz physical (PHY) and medium access control (MAC) for short-range unlicensed communications. The standard provides up to 10 Gb/s wireless personal area network (including point-to-point) transport for both bulk data transfer and multimedia streaming. TG20 is considering three device types; ranging from high-end devices with steerable antennas to low-end devices for cost effective, short range, gigabit solutions. This underlines the role of the millimetre wave antenna in gigabit communications.

Table 1.6 summarises potential applications of millimetre wavelength systems as submitted in response to the IEEE Call for Applications (CFA). The submissions illustrate the support for some of the applications listed. The applications have been arranged in the numeric order of the IEEE CFA document number (last column)[34].

Table 1.6 Possible applications for millimetre wave communications. (Reproduced by permission of © 2007 IEEE [34])

No.	Description of applications	Outdoor	Indoor	IEEE CFA Doc. number
1	Gigabit Ethernet link, wireless IEEE1394 applications	–	• LOS • Data rate: $\leq$ 1 Gb/s duplex • Range: $\leq$ 17 m	04-0019
2	Ad hoc information distribution system	–	• LOS • Data rate: 622 Mb/s • Range: $\geq$ 20 m (AP-AP) $\geq$ 3 m (AP-MT)	04-0097
3	Multimedia, information distribution system	–	• LOS • Data rate: $\geq$ 1 Gb/s • Range: $\leq$ 10 m	04-0098
4	• Outdoor: fixed wireless access, distribution in stadiums, intervehicle communication, etc. • Indoor: connecting multimedia devices (wireless home link), ad hoc meeting, heavy content download, distribution system	• LOS • P2P, P2MP • Data rate: 156 Mb/s to 1.5 Gb/s • Range: 400 m to 1 km	• LOS • Data rate: 100 Mb/s to 1.6 Gb/s • Range: ~10 m	04-0118
5	Small office/meeting scenario, general office applications	–	• NLOS • OFDM • Data rate: $\leq$ 200 Mb/s • Range: 2 to 4 m	04-0141
6	Distribution links in apartments, stadium, etc.	• LOS • P2P • Bandwidth: > 300 MHz • Range: $\leq$ 220 m	–	04-0153
7	Wireless home video server connected to HDTV, PC and other video devices	–	• LOS • Data rate: 300 Mb/s, 400 Mb/s and 1.5 Gb/s uncompressed HDTV data • Range: $\leq$ 10 m	04-0348

(continued overleaf)

Table 1.6 (*continued*)

No.	Description of applications	Outdoor	Indoor	IEEE CFA Doc. number
8	• Outdoor: distribution links in apartments, stadium, etc. • Indoor: ad hoc network	• LOS • P2P and P2MP • Bandwidth: > 300 MHz • Range: ≤ 220 m	• LOS • Data rate: ≥ 1 Gb/s and ≥ 622 Mb/s • Range: ≥ 20 m and ≥ 3 m	04-0352
9	PowerPoint and such applications	–	• LOS and NLOS • Data rate: ≥ 1 Gb/s • Range: ≤ 3 m • Space diversity	04-0514
10	• Replacement for 1394 FireWire • Replacement for USB • Military – future combat systems, secure communication	–	• LOS and NLOS (people) • 100 to 500 Mb/s link, 1 Gb/s in 2007 • Short range	04-0665

1.3 Millimetre Wave Characterisations

This section presents benefits of 60 GHz technology and its major characteristics. It can be used for high-speed Internet, data and voice communications, and offers the following key benefits:

1. Unlicensed operation
2. Highly secure operation: resulting from short transmission distances due to oxygen absorption, narrow antenna beamwidth and no wall penetration
3. Virtually interference-free operation: resulting from short transmission distances due to oxygen absorption, narrow antenna beam width and limited use of 60 GHz spectrum
4. High level of frequency re-use enabled: the communication needs of multiple customers within a small geographic region can be satisfied
5. Fibre optic data transmission speeds possible: 7 GHz (in the USA) of continuous bandwidth available compared to < 0.3 GHz at the other unlicensed bands (3.5 GHz internationally available)
6. Mature technology: long history of this spectrum being used for secure communications
7. Carrier-class communication links enabled: 60 GHz links can be engineered to deliver "five nines" (99.999 %) availability if desired (outdoor applications such as backbone or bypass bridges)

There is a widespread belief that the characteristics of a millimetre wave present many difficulties in terms of propagation environment for high data rate wireless communications.

While the oxygen absorption does indeed cause a 15 dB/km loss, this translates to only a 1.5 dB loss at 100 m, so for indoor applications the absorption loss from oxygen is small, if not negligible.

Another loss – proportional to the frequency squared – comes from the *Friis path loss* equation (1.2). This "loss", however, can be attributed to another factor. If omni-directional antennas, such as half-wavelength dipoles, are used, then as the frequency rises, the effective area of the antennas decreases as frequency squared. If, on the other hand, the (physical) area of the antennas is kept constant, then there is no increase in path loss because the electrical area increases as the wavelength decreases (squared).

For instance, a 60 GHz antenna, which has an effective area of 1 square inch, will have a gain of approximately 25 dBi, but this gain comes at the expense of being highly directional. This would mean that for millimetre wave radios to be used at their full potential they would need a solution for precise pointing.

1.3.1 Free Space Propagation

As with all propagating electromagnetic waves, for millimetre waves in free space the power flux density falls off as the square of range. For a doubling of range, power flux density at a receiver antenna is reduced by a factor of four. This effect is due to the spherical spreading of the radio waves as they propagate. The frequency and distance dependence of the loss between two isotropic antennas can be expressed in absolute numbers by (in dB):

$$L_{\text{free space}} = 20 \ \log_{10} \left(4\pi \frac{R}{\lambda} \right) (\text{dB}) \qquad (1.1)$$

where $L_{\text{free space}}$ is the freespace loss, R is the distance between transmit and receive antennas, and λ is the operating wavelength. This equation describes line-of-sight wave propagation in free space. This equation shows that the free space loss increases when the frequency or range increases. Thus, millimetre wave free space loss can be quite high, even for short distances. This indicates that the millimetrewave spectrum is best used for short-distance communications links. The Friis equation (1946) gives a more complete expression for all the factors from the transmitter to the receiver (as a ratio, linear units) [35]:

$$P_{Rx} = P_{Tx} G_{Rx} G_{Tx} \frac{\lambda^2}{(4\pi R)^2 L} \qquad (1.2)$$

where G_{TX} = transmitter antenna gain, G_{RX} = receiver antenna gain, λ = wavelength (in the same units as R), R = line-of-sight (LOS) distance separating transmit and receive antennas and L = system loss factor (≥ 1).

1.3.2 Millimetre Wave Propagation Loss Factors

In microwave systems, transmission loss is accounted for principally by the free space loss. However, in the millimetrewave bands additional (absorption) loss factors come into play, such as gaseous losses and rain (or other micrometeors) in the transmission medium. Factors that affect millimetre wave propagation are given in Figure 1.6.

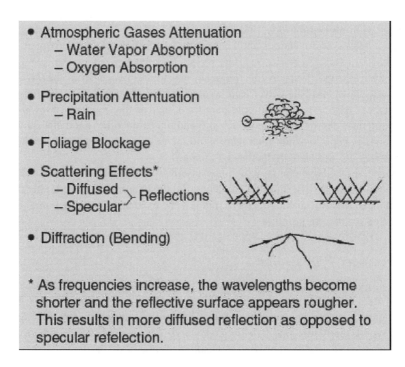

Figure 1.6 Propagation effects influencing millimetre wave propagations. (Reproduced by permission of © 2005 IEEE [36])

1.3.3 Atmospheric Losses

Transmission losses occur when millimetre waves travelling through the atmosphere are absorbed by molecules of oxygen, water vapour and other gaseous atmospheric constituents. These losses are greater at certain frequencies, coinciding with the mechanical resonant frequencies of the gas molecules.

The H_2O and O_2 resonances have been studied extensively for the purpose of predicting millimetre wave propagation characteristics. Figure 1.7 shows an expanded plot of the atmospheric absorption versus frequency at altitudes of 4 km and sea level, for water content of 1 and 7.5 gm/m³ respectively (the former value represents relatively dry air while the latter value represents 75 % humidity for a temperature of 10°C).

1.4 Channel Performance

Planning for millimetre wave spectrum use is based on the propagation characteristics and channel performance of radio signals and the noise apparent in this frequency range. While signals at lower frequency bands, such as a GSM signal, can propagate for many kilometres and penetrate more easily through buildings, millimetre wave signals can travel only a few kilometres or less, and suffer from high transmission loss in the air and solid materials. However, these characteristics of millimetre wave propagation can be very advantageous in some applications.

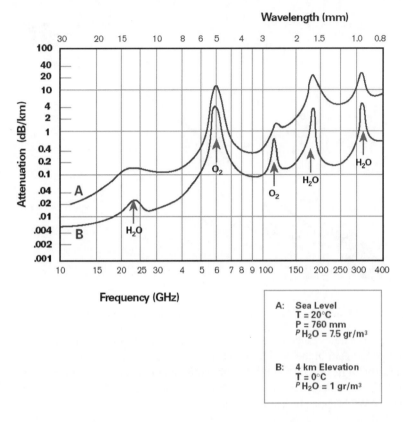

Figure 1.7 Average atmospheric absorption of millimetre waves. (Reproduced by permission of © 2005 IEEE [36])

Millimetre waves can establish more densely packed communications links, thus providing very efficient spectrum utilization; the high absorption enabling shorter range frequency re-use, and therefore increasing the overall capacity of communication systems. The characteristics of millimetre wave propagation are summarised in this section, including free space propagation and the effects of various physical factors on propagation.

The main challenges for a 60 GHz channel can be described as follows:

- High loss from the Friis equation
- Doppler shift is non-negligible at pedestrian velocities
- Human shadowing
- Non-line-of-sight propagation, which induces random fluctuations in the signal level, known as multipath fading, as shown in Figure 1.8
- Noise

The transmitting power of a 60 GHz communications link is restricted to +40 dBm EIRP limit by the FCC in the USA. Transmitter power and path loss can be limiting factors

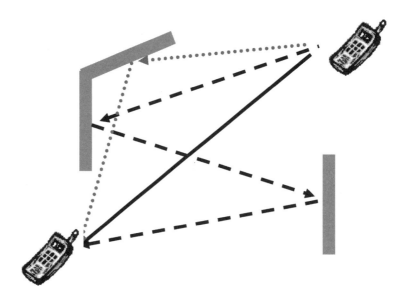

Figure 1.8 Multipath effect for indoor wireless communications

for a high-speed wireless link. However, at these frequencies antenna directivity can be used to increase power gain in the desired direction.

The capacity limits of a 60 GHz link with omnidirectional antennas at both ends should be considered. Even when the bandwidth is unlimited, the received power P_{Rx} is still limited by the Shannon AWGN capacity, as given by:

$$C = BW \ \log_2 \left(1 + \frac{P_{Rx}}{BW \ N_o} \right) \approx 1.44 \frac{P_{Rx}}{N_o} \quad \text{when} \quad BW \to \infty \qquad (1.3)$$

The result is shown in Figure 1.9. As can be seen, it is very unlikely that an omnidirectional antenna can be used to achieve a Gb/s data rate when human shadowing exists. When the transceiver has $P_{Tx} = 10$ dBm, $NF_{Rx} = 10$ dB and the environment has a human shadowing loss of 18 dB, α needs to be in the range of 10 to 15 dB for 1 Gb/s at 60 GHz; the results for other values of α are shown in Reference [38]. This means that the total antenna gain has to be approximately at least 30 dB.

Ignoring the human shadowing loss, means that there exists a clear path between the transmitter and receiver. A 60 GHz system with the following parameters can be considered as an illustration:

Tx power, P_{Tx} 10 dBm
Noise figure, NF 6 dB
Implementation loss, IL 6dB
Thermal noise, N 174 dBm/MHz
Bandwidth, B 1.5 GHz
Distance, R 20 m
Path loss at 1 m, PL_0 57.5 dB

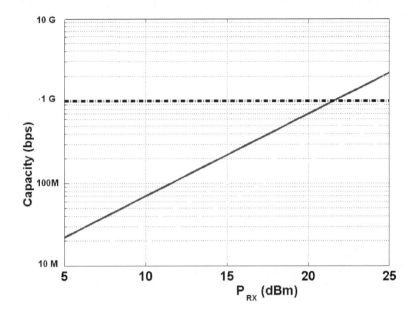

Figure 1.9 Shannon limit with distances $d = 10$ m between a transmitting omnidirectional antenna and a receiving omnidirectional antenna [37]

the ratio of signal power to noise power (SNR) at the receiver can be calculated, as (in dB):

$$\text{SNR} = P_{Tx} + G_{Tx} + G_{Rx} - PL_0 - PL(R) - IL - [KT + 10\log_{10}(B) - NF] \quad (1.4)$$

where G_{Tx} and G_{Rx} denote the transmit and received antenna gain respectively. P_{Tx} denotes transmitter power, PL_0 is the path loss at 1 m and B is the bandwidth, and the link length is R. Inserting Equation (1.4) into the Shannon capacity formula of Equation (1.3), the maximum achievable capacity in an AWGN can be calculated. In non line of sight (NLOS) links the path loss due to scattering exceeds the square law for free space links. This path loss exponent can vary from 2 (LOS) to 5 in extreme NLOS links. The path loss exponent n is more fully explained in Reference [39]. Figure 1.10 shows the Shannon capacity limit for an indoor office in the LOS and non-LOS (NLOS) cases, using an omni-directional antenna configuration. It can be observed that for the LOS condition, a 5 Gb/s data rate is not possible at any distance. Whereas, the operating distance for the NLOS condition is limited to below 3 m, though the capacity for NLOS decreases more drastically as a function of distance.

To improve the capacity for a given operating distance, either the bandwidth or signal-to-noise ratio (SNR) or both should be increased. It can also be seen from Figure 1.10, that increasing the bandwidth used by more than 4 times only significantly improves the capacity for distances below 5 m. Beyond this distance, the capacity for the 7 GHz bandwidth is only slightly above the case of the 1.5 GHz bandwidth, since the SNR at the Rx is reduced considerably at longer distances due to higher path loss. But, the overall capacity over the considered distance increases notably if a 10 dBi transmit antenna gain is employed, as compared to the omnidirectional antenna for both 1.5 and 7 GHz bandwidths. This clearly shows the importance of antenna gain in providing a very high data rate application at 60 GHz, which it is not possible

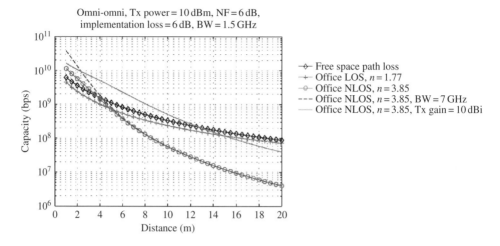

Figure 1.10   Shannon capacity limits for the case of an indoor office using the omni-omni antenna setup. (Reproduced by permission of © 2007 S. K. Yong and C.-C. Chong [6])

to provide with the omni-directional antenna configuration. However, this does indicate how much gain is required.

The capacity as a function of combined Tx and Rx gain for an operating distance of 20 m is plotted in Figure 1.11. To achieve 5 Gb/s at 20 m, a combined gain of 25 and 37 dBi are indicated for LOS and NLOS, respectively, with no shadowing. This is a practical value since it is a combined Tx and Rx gain. However, to achieve the same data rates in multipath channels, a higher gain is needed to overcome the fading margin.

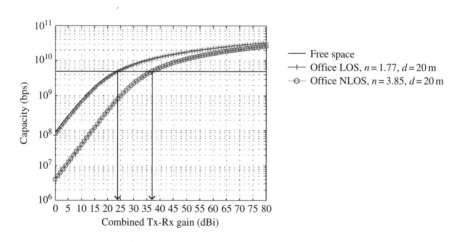

Figure 1.11   The required combined Tx–Rx antenna gain to achieve a target capacity. (Reproduced by permission of © 2007 S. K. Yong and C.-C. Chong [6])

Because directional antennas are required for gigabit wireless communications, there can be different configurations for the access point (AP) and mobile terminal (MT) depending on the application, as shown in Figure 1.12.

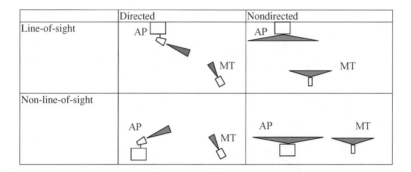

Figure 1.12 Classification of millimetre wave links according to the antenna beamwidth of the access point (AP) and mobile terminal (MT), in respect of the existence of a line-of-sight path. The radiation beamwidth is shown in grey

Consider a 60 GHz measurement as shown in Figure 1.13. The synthesiser has a maximum output power of 0 dBm (1 mW) at 65 GHz. The connecting coaxial cables have a transmission loss of a maximum of 6.2 dB/m at 60 GHz. The conversion loss of the subharmonic mixer is assumed to be 40 dB and its noise figure is 40 dB, while the voltage standing wave ratio (VSWR) is 2.6:1. The noise floor for the spectrum analyser is assumed to be −130 dBm.

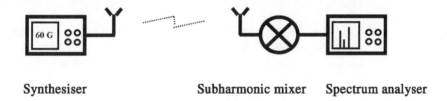

Synthesiser   Subharmonic mixer   Spectrum analyser

Figure 1.13 Channel measurement setup

The dynamic range for this configuration can be measured as a function of the total antenna gain and the separation of the antennas at 60 GHz. The result is shown in Figure 1.14.

Multipath propagation occurs when waves emitted by the transmitter travel along a multiplicity of different paths and interfere with waves travelling in a direct line-of-sight path. Fading is caused by the destructive interference of these waves. This phenomenon occurs because waves travelling along different paths may be out of phase when they reach the antenna, thereby cancelling each other to form an electric field null. Since signal cancellation is almost never complete, one method of overcoming this problem is to transmit more power (either omnidirectionally or directionally). In an indoor environment, multipath propagation is almost always present and tends to be dynamic (constantly varying) due to moving scatterers. Severe fading due to the multipath can result in a signal reduction of more than 30 dB. It is

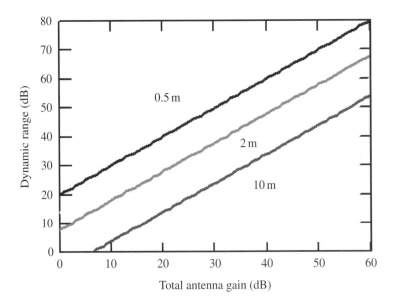

Figure 1.14 Dynamic range as a function of total antenna gain and distance between the antennas at 60 GHz

therefore essential to provide an adequate link margin to overcome this loss when designing a wireless system. Failure to do so will adversely affect the reliability of the link. The amount of extra RF power radiated to overcome this phenomenon is referred to as a fade margin. The exact amount of fade margin required depends on the desired reliability of the link, but a good estimate is 20 to 30 dB.

In channel measurements, as shown in Figure 1.15, antennas with different beamwidths are compared. For antennas with a narrow beamwidth, a notch appears in the frequency response. For antennas with a broad beamwidth (many multipaths received), the notch in the frequency response becomes severe. In the extreme case, if the antenna beam is as narrow as a laser, this notch will not exist in the frequency response.

The notch width is affected by the range of delays (delay spread), while the notch depth is affected by the difference in path gain (or loss) for the multipath signals. In addition, the notch position in the frequency domain is affected by the length differences between the propagation paths.

To minimise the notch effect, a number of solutions can be considered. One is to employ a narrow-beam antenna to reduce reflected paths and achieve a smaller notch depth (fewer multipaths). However, the problem of tracking resolution and the speed of tracking (pointing) of the narrow beam antenna will need to be solved. Alternatively, precise source tracking or space diversity can be used to avoid the notch effect. However, there are some issues that still need to be tackled in multi-antenna implementations.

In an office environment, reflection characteristics of interior structures have been studied and reported in [40]. Human shadowing was investigated and typical results are summarised in Figure 1.16. When 0 dBm power at 60 GHz is transmitted via a 10 dBi gain transmitting antenna to a receiving antenna with 10 dBi gain at a distance of 4 m, the spectrum

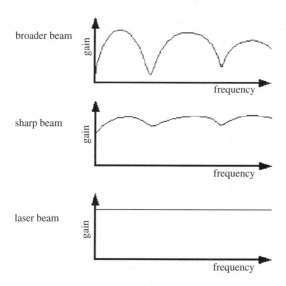

Figure 1.15   Beam width and channel distortion

shows that the received power is −35 dBm approximately when there is no human shadowing. (Case 1). If there is a human body between two antennas, the signal is reduced to the range of between −55 and −65 dBm (Case 2). If there are two human bodies between two antennas, the signal is reduced to the range of between −65 and −80 dBm (Case 3). If the beam direction of the 10 dBi transmitting antenna is changed so that the signal can bounce off a concrete ceiling at a height of 2 m and be reflected to the receiver, the received signal is

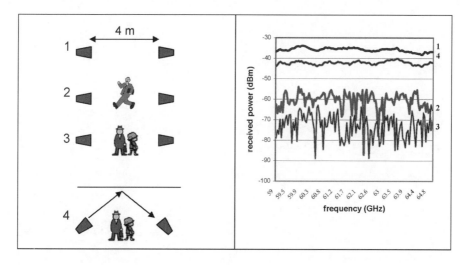

Figure 1.16   Indoor channel measurement at 60 GHz for NLOS. Transmitting antennas and receiving antennas have a 10 dBi gain. Case 1 shows a line-of-sight scenario. Case 2 shows a person standing between two antennas. Case 3 shows two people standing between two antennas. Case 4 shows a non-line-of-sight wireless link [40]

increased to $-42\,$dBm (Case 4). This illustrates that reflected propagation at 60 GHz can be used for non-line-of-sight wireless communications.

1.5 System Design and Performance

Cost-effective millimetre wave solutions for high data rate transmissions at 60 GHz still need to be determined. In this respect, some important selections have to be made which might be crucial for its commercial success:

- Selection of antennas
- Selection of the 60 GHz radio front-end architecture

1.5.1 Antenna Arrays

A presumed advantage of a 60 GHz radio is the small antenna area compared to a lower-frequency wireless system. Thus, it becomes possible to integrate antenna arrays into portable devices, and the antenna directivities can be improved. While it is possible to increase the antenna gain for a single antenna (e.g. using mechanical structures such as a horn antenna), it is more desirable to increase the directivity by employing an antenna array or multiple-input multiple-output system as shown in Figure 1.17. For a fixed antenna aperture size A the directivity is $D = 4\pi a/\lambda^2$, and from Equation (1.2) it can be seen that there is actually an improvement in the received power by moving to higher frequencies for a fixed antenna form factor. For example, a 60 GHz system with a 16-element antenna array has a 3 dB gain over a 5 GHz omnidirectional system while occupying only 10 % of the antenna area.

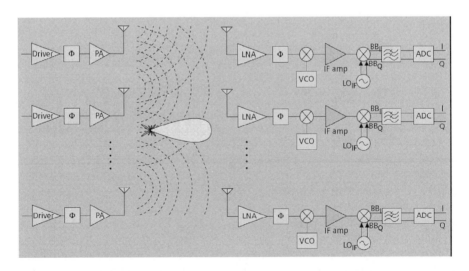

Figure 1.17 A generic multiple transceiver architecture with beam-steering antennas. (Reproduced by permission of © 2004 IEEE [41])

1.5.2 Transceiver Architecture

A generic adaptive beamforming multiple antenna radio system is shown in Figure 1.17. It is assumed that the antenna elements are small enough to be directly integrated into the package or potentially even on-chip. The main benefit of the multiantenna architecture used here is the increased gain that the directional antenna pattern can provide, which as has been seen, is needed in order to support multigigabit per second data rates at typical indoor distances. In addition to the antenna gain, the use of antenna arrays also provides spatial (or angular) diversity, automatic spatial power combining, and an electronic beam steering function. The transceiver architecture in Figure 1.17 depicts N independent transmit and receive chains. Such an approach would enable a flexible multiple-input multiple-output (MIMO) system that could fully exploit a multipath-rich environment for increased capacity and/or robustness [41].

The main disadvantage with this arrangement is the high transceiver complexity and power consumption since there is little sharing of the hardware components. Measurements of the 60 GHz channel properties indicate that most of the received energy is contained in the specular path [42], so a full MIMO solution targeting capacity may not be able to benefit fully from this channel. A more efficient implementation would be to use a phased array that takes the identical RF signal and shifts the phase for each antenna to achieve beam steering. Essentially, communication systems can select one strong path and apply an angular or spatial filter, forming a narrow beam in the direction of the chosen signal [43]. This approach significantly reduces hardware costs, as most of the transceiver can be shared with the addition of controllable phase shifters between the transceiver and antenna array.

For the choice of the architecture of the 60 GHz front-end radio there are, in principle, four options:

1. Employing superheterodyning architecture
2. Employing direct conversion architecture
3. Employing five-port technology
4. Employing software radio architecture

1.5.2.1 Superheterodyning Architecture

With regard to the superheterodyning option, a simple architecture is considered as depicted in Figure 1.18(a). This figure shows a basic 60 GHz RF front-end architecture for application at the portable station (PS) end. Ideally it should be an integrated on-chip solution consisting of a receive branch, a transmit branch and a frequency generation function. The receive branch consists of the receive antenna, a low-noise amplifier (LNA) and a mixer that downconverts to IF. The transmit branch consists of a mixer, a power amplifier (PA) and the transmit antenna. The antennas are (integrated) patch antennas. The mixers are image rejecting mixers (they do not need to be in-phase/quadrature (IQ) mixers). The IF in this example is taken as 5 GHz with the idea that, with appropriate modifications, an IEEE 802.11a RF chip set can serve as the IF here, to allow dual-mode operation and interoperability. Superheterodyning architecture requires more components and more DC power so is unsuitable for mobile devices.

1.5.2.2 Direct Conversion Architecture

The advantages of a direct conversion are that it is well suited to monolithic integration, due to the lack of image filtering and its intrinsically simple architecture [44, 45]. FSK modulated signals are especially well-suited to direct conversion, due to their low-signal energy at DC. However, the direct conversion receiver has not gained widespread acceptance to date, especially in high-performance wireless transceivers, due to its intrinsic sensitivity to DC offset problems, harmonics of the input signal and local oscillator (LO) coupling problems back in to the antenna. Offset arises from three sources [46]:

1. Transistor impedence mismatch in the signal path
2. LO signal leaking to the antenna because of poor reverse isolation through the mixer and RF amplifier, and then reflecting at the antenna terminals and ultimately self-downconverting to DC through the mixer
3. Strong adjacent or near channel signal leaking into the LO part of the mixer, which then self-downconverts to DC

Good circuit design may reduce these effects to a certain extent, but they cannot be eliminated completely, particularly so if quadrature phase shift keying (QPSK) or Gaussian minimum shift keying is used since the spectra of these schemes possess a peak at DC. However, when orthogonal frequency division multiplexing (OFDM) is used there may be a solution, which avoids the use of those subcarriers which, after conversion, correspond with, or will be close to, the DC component. There may also be other solutions that exploit the particularities of the 60 GHz physical layer.

A block diagram of an example millimetre wave direct conversion architecture is shown in Figure 1.18 (b). This example consists of transmit and receive paths which combine with a 60 GHz switch at the antenna side.

The voltage-controlled oscillator (VCO) operates in the 3–4 GHz range. This VCO is modulated with the data stream (>1 Gb/s), which does not affect the low bandwidth phase-locked loop (PLL) circuitry. The modulated signal is multiplied (16 times for the transmit side and 8 times for the receive paths) and filtered, before being transmitted or used to drive the subharmonic receiver mixer.

To support output power requirements, two amplifier monolithic microwave integrated circuits (MMICs) are cascaded in series. A low-noise amplifier (LNA) in the receive chain guarantees low-noise figure values. The most important issues for the functionality of the architecture are the filters placed after each multiplier stage. Each filter must be designed to avoid unwanted emissions in the transmit and receive bands.

The voltage-controlled oscillator (VCO) can be driven by an (off-chip) frequency synthesizer. In conventional designs the VCO is usually implemented off-chip because it occupies too much area on the chip without providing sufficient performance. At frequencies as high as 60 GHz it may become, however, feasible to implement the VCO directly on the chip because the minimum dimensions to achieve the required performance become much smaller. The advantage of this approach is the reduction in components that have to be mounted on an external circuit board and the avoiding of on-chip frequency multiplier circuits, thus saving space on the chip and reducing any VCO performance degradation that could arise. It is important to note that an on-chip VCO, that directly generates a reference frequency close to 60 GHz,

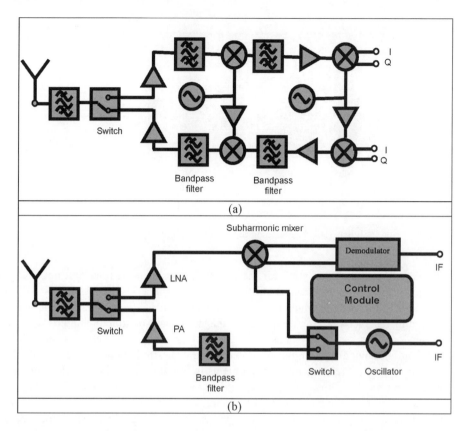

Figure 1.18 (a) Block diagram for millimetre wave/microwave circuits. (b) Block diagram for a 60 GHz direct conversion architecture

may have a relatively lower performance when compared with the requirements of a VCO that operates on a much lower frequency in combination with a couple of frequency multipliers.

1.5.2.3 Five-Port Radio

The five-port technology (or six-port technology), described in [47] is a passive linear device, composed of two input ports and three outputs (see Figure 1.19). A phase shifter is used to adjust the phase between RF and LO. On the ports of P1, P2 and P3, diode detectors are used in each port, instead of mixers, as the frequency converter. Five-port technology has been extended to direct digital transmitters and can be used for software-defined radio applications, as it can accommodate different wireless modulation standards without requiring hardware modification.

1.5.2.4 Software Defined Radio

Employing analogue-to-digital conversion (ADC) and digital-to-analogue conversion (DAC) directly at the antennas would appear to make the complete RF and IF part of the transceiver

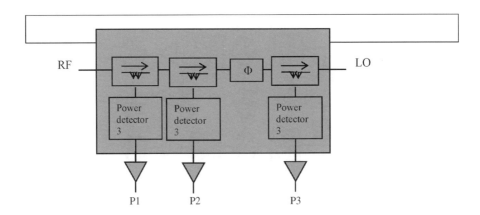

Figure 1.19   Block diagram for the five-port technology

chain obsolete. However, this option for the current purpose can be ruled out immediately because this would require ADC and DAC devices operating at a 60 GHz or more. A low-cost implementation of this in the medium term is not considered feasible. An alternative approach, the subsampling receiver, is claimed to represent the "ultimate" solution for simple low-power downconversion. This essentially consists of a sampling switch, clocked at a much lower frequency, and an analogue-to-digital (A/D) converter. The limitations of the subsampling approach, however, illustrate the inherent problems in low-power receiver implementations. In a subsampling receiver, image frequencies exist at integral multiples of the sampling rate and can alias (map) onto the band of interest. As a result, careful filtering prior to the down-conversion is required. For example, downconversion of an RF signal having a bandwidth of 500 MHz would require a sampling rate of at least 1 GHz, assuming a "brick wall filter" (a filter with infinite cut off outside the working band). In practice, the sampling rate would have to be much higher – at least 2 GHz – in order to minimise the effects of the filter. It is questionable whether a 2 GHz ADC with a 10 bit quantisation, will become available in the medium term. In addition, the signal-to-noise ratio (SNR) of the downsampled signal will inevitably be poorer than that of an equivalent system employing a mixer for downconversion. This is due to the noise aliased from the bands between DC and the passband [48].

1.6 Antenna Requirements

In this book, an overview is presented, of an approach within the application area that utilises millimetre wave antenna technology and offers significant promise in making these Gb/s wireless links a reality. Throughout the book, topics will be revisited and approached from different angles in order to present alternative ways of analysing the various components and parameters that make up millimetre wave systems.

For a single antenna element with an antenna gain of more than 30 dBi with a half-power beamwidth (HPBW) of approximately 6.5°, a reliable communication link is difficult to establish even in a LOS condition at 60 GHz. This, as has been seen, can be due to human movement which can easily block and attenuate such a narrow beam signal. To overcome this problem, a switched beam antenna array or adaptive antenna array can be implemented to search and beamform, in order to capture the available signal. The array is required to track the signal path either continuously or periodically, depending on the stability of the link. One major parameter of the performance of the link is how many antenna elements are required to achieve the intended antenna gain. This is a separate consideration from the array gain, which refers to the performance improvement in terms of the SNR over a single antenna element. Also of interest is the angular resolution or beamwidth of such antennas, since this defines the number of multipaths that the antenna sees in a scattering environment. The directivity of the linear array is given by [49]:

$$D = \frac{4\pi}{\iint |F_n(\phi, \theta)|^2 \sin \theta \, d\theta \, d\phi} \quad (1.5)$$

where $F_n(\varphi, \theta)$ is the normalized field pattern, which can be expressed as a product of the normalized element pattern and the normalized array factor. The variables φ and θ represent the azimuth and elevation angle, respectively. For a uniform linear array, the normalized array factor can be expressed as:

$$f_n(\varphi, \theta) = \frac{\sin[(N/2)(kd \cos \theta + \beta)]}{N \sin[(1/2)(kd \cos \theta + \beta)]} \quad (1.6)$$

where N, d and β are the number of antenna elements, the antenna spacing between adjacent elements and the phase shift between elements, respectively. For an omnidirectional antenna, it can be shown that up to 100 omni-element arrays are required to achieve a gain of only 23 dBi, which is far from the requirement discussed previously. Hence a more directive/higher gain element is required to improve the overall gain of the array.

Many types of antenna structures are considered not suitable for 60 GHz WPAN/WLAN applications due to the requirements for low cost, small size, light weight and high gain. In addition, 60 GHz antennas are also required to be operated with approximately constant gain and high efficiency over the broad frequency range (57–66 GHz). The importance of beamforming at 60 GHz has been introduced in Section 1.4, and can be achieved by either switched beam arrays or phased arrays. Switched beam arrays have multiple fixed beams that can be selected to cover a given service area. They can be implemented more easily compared to phased arrays, which require the capability of continuously varying a progressive phase shift between the elements.

The complexity of phase arrays at 60 GHz typically limits the number of elements. In Reference [50], a 2×2 beam-steering antenna with circular polarization at 61 GHz was developed. The gain is approximately 14 dBi with 20° half power beamwidth (HPBW). Similarly, in Reference [51], another 60 GHz integrated four-element planar array was developed. Each antenna is integrated with a subharmonic I/Q mixer for the convenience of high-speed signal processing, such as adaptive beamforming. The implementation of a larger phased array, however, presents technical challenges, such as the requirement for a higher feed network loss, a more complex phase control network, stronger coupling between antennas as well as feedlines, etc. These challenges make the design and fabrication of larger phase arrays more complex and expensive. Hence, research is required to develop a low-cost, small-size, light-weight and high-gain steerable antenna array that can be integrated into the RF front-end electronics.

To achieve this, the design approach can be focused on either:

(a) accepting the presence of multipath (with delays corresponding to the room size) and mitigating it with equalisation techniques or
(b) using line-of-sight links with narrow-beam antennas to eliminate virtually all multipaths, and thus use simple unequalised modulation schemes, such as FSK and PSK.

In the first case, the design effort would concentrate on narrow beam antenna design techniques, whereas in the second approach, the work would concentrate on antenna/beam-steering techniques. These must be used because multipath delay in the typical indoor environment is on the order of the target bit period (tens of nanoseconds) and causes intersymbol interference. The multipath delays for indoor systems depend on the size of a room and the density and placing of scatterers within the illuminated space.

It is assumed for the moment that for high speed data transmission a simple two- or four-level FSK or PSK system is used, because complex modulation schemes such as equalisation, diversity or multicarrier techniques are deemed to be impractical or too expensive for 60 GHz. For such a simple system to work reliably the channel impulse response should not contain significant multipath components, so that the data rate is not limited by multipath effects. Also an initial assumption is made, that high-speed and high-capacity WLANs can use a femtocellular architecture, with a single cell for each room and multiple cells for a large open area office.

For the "LOS with narrow beam antennas" approach, the amount of multipath power will depend on the number of paths between the transmitter and receiver, which in turn will depend on the directivity of the antennas at the transmitter and receiver, as well as specific environmental factors. It will also depend on the ability of the antenna to resolve the multipaths' angular space. If omnidirectional antennas are used at both the transmitter and receiver, then there will be many possible paths, whereas if highly directional antennas are used, there may be only a single LOS path. Once the beamwidth is sufficiently narrow, there is no significant multipath in most practical circumstances. (Of course, if the transmit and receive LOS is perpendicular to a pair of parallel reflectors an infinite number of multipaths will occur.)

To explore the consequences of this approach, three different antenna designs are now considered.

1. Phased Array

Considering an 8×8 phased array antenna with beam steering, this arrangement requires complex phase shifters (or hybrid Tees and attenuators applied to the I and Q channels), and therefore is subject to high loss at 60 GHz. These losses reduce the effective gain of the antenna array. In addition, currently there is no phase shifter MMIC available at 60 GHz on the market so a hybrid Tee and real weights (attenuators) would be needed to build phase- shifting functions. In addition, the beam shape becomes asymmetric when the beam direction moves away from the z axis (this is generally called aberration). This means that the sidelobes of the radiation pattern will grow when the beam is away from broadside. Also, circular polarisation at wide angles with a phase-shifted array is almost impossible to achieve. The performance of circular polarisation is unlikely to be achieved as the phased array becomes increasingly complex. The main challenge of this design is to have a complex phase shifter and to have low loss. Lastly, it is also difficult to achieve good circular polarisation in all directions.

2. A 2 × 2 Horn Array Plus Beam Switching

The gain of this design is limited by the size and the separation distance of horns. Each unit consisting of a 2 × 2 element array acts as an independent source [52]. This design requires a multibit phase shifter but generates good circular polarisation. By adding several tilted horns, this design can have ±100° coverage. The feeding network needs to have the correct amplitude and phase in the two orthogonal linear polarisations in order to generate good circular polarisation. The main challenge of this design is to reduce the sidelobe level caused by using 2 × 2 elements.

3. Beam Switching Array

This design uses a minimum number of elements (4 × 4) to achieve ±100° coverage [53]. No phase shifter is required. Each element generates an independent beam. The configuration operates in a different manner to that of the phased array, and there is no size limit for each element. Each element can be optimised individually to meet the specifications for the individual links. Sidelobe levels can therefore be controlled by a single horn design. The gain of each element can be improved by adding a superstrate together with a horn, or using stacked patches. More details about gain enhancement can be found in Chapter 2. The feed network needs to have the correct amplitude excitation for each element, but not the phase. Circular polarisation can be improved by a tilted waveguide or helical element. More details of this configuration will be discussed in Chapter 4. The main defining parameters for these designs are compared in Table 1.7.

Table 1.7 Comparison of three 60 GHz antenna designs

	8 × 8 phased array	2 × 2 array plus beam switching	16 beam switching array
High gain	Yes (but the loss of phase shifters is also high)	Yes	Yes
HPBW 20°	Yes	Yes	Yes
Sidelobe −10 to −20 dB	Not at the 100° beam direction	Not easy	Yes
Circular polarisation	Medium	Possible	Possible
Beam steering range	Beam direction is controlled by phase shifting. Sidelobe level increases when the beam is away from broadside.	Beam direction is controlled by the height of horns and phase shifters.	Beam direction is controlled by switches.
Feeding point design	Amplitude, phase	Amplitude, phase	Amplitude
Phase shifters	Complex	2 bits	No
Challenge	Complex phase shifter, low-loss phase shifter	Sidelobe reduction	High gain with small size

1.7 Link Budget

The link budget is used to determine system capabilities under a range of operating conditions for the specified data rates, ranges and bit error rate. The expressions below identify the necessary parameters, which can be used to calculate the final link margin:

Path loss at 1 m ($PL_0 = 20 \log_{10}(4\pi f_c/c)) = 68.00$ dB
   where f_c (centre frequency) $= 60$ GHz, $c = 3 \times 10^8$ m/s
Average noise power per bit (dB) $= N = -174 + 10 \times \log_{10}(R_b)$
   where R_b (Gb/s) is the bit rate
Average noise power per bit (dBm) $P_N = N + $ Rx noise figure (referred to the antenna
   terminal) (dB)
Total path loss (dB) $= PL = P_T + G_T + G_R - P_N - S - M_{\text{shadowing}} - I - PL_0$

where P_T is average Tx power (dBm)
          G_T is Tx antenna gain (dBi)
          G_R is Rx antenna gain (dBi)
          S is minimum E_b/N_0 for the AWGN channel (dB)
          $M_{\text{shadowing}}$ is shadowing link margin (dB)
          I is implementation loss (dB), including filter distortion, phase noise, frequency
            errors
Maximum operating range $d = 10^{PL/10n}$ (m)

   where n is path loss exponent, subject to the scenario.

The following path loss parameters are considered by the IEEE 802.15.3c standard [54]:
For LOS scenarios:

- Path loss at 1 m: $PL_0 = 68$ dB
- Path loss exponent: $n = 2$
- Shadowing link margin: $M_{\text{shadowing}} = 1$ dB

For NLOS scenarios:

- Path loss at 1 m: $PL_0 = 68$ dB
- Path loss exponent: $n = 2.5$
- Shadowing link margin: $M_{\text{shadowing}} = 5$ dB

A simple millimetre wave link can be represented as in Figure 1.20.

From this perspective the signal-to-noise ratio for the system can be calculated. In Table 1.8, an example such as the configuration in Figure 1.16 (Case 4) is used to calculate the signal and noise of a millimetre wave system with two 15 dBi directional antennas in a 5 m wireless link. Both free space loss, and reflection loss are taken into account.

When the transmitted signal level is set to 15 dBm, a 5 dB loss can be expected due to the feeding network of transmitting antennas. The power delivered to the antenna is therefore 10 dBm. The EIRP is then effectively increased to 25 dBm when the transmitting antenna has a

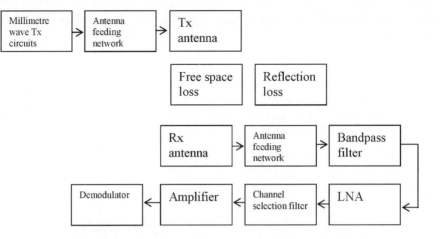

Figure 1.20   The 60 GHz transmitter, receiver and wireless link

Table 1.8   An example of a millimetre wave scenario

Transmit power (dBm)	10
Bandwidth (GHz)	2
Distance (m)	5
Free space loss (dB)	81.98419713
Tx antenna gain (dBi)	15
Rx antenna gain (dBi)	15
Reflection loss (dB)	15
Input level (dBm)	−56.98419713
Input noise level (dBm)	−81

15 dBi gain. In a 5 m link, the 60 GHz signal suffers a free space loss of approximately 81.98 dB. The signal is therefore attenuated by 81.98 dB due to this free space loss, and a further 15 dB due to reflection loss. The final EIRP at the receiving antenna is therefore −72 dBm.

The input noise of the converter is the theoretical thermal noise floor limit, *KTB*. *KTB* is calculated as follows:

$$KTB = 4.002 \times 10^{-21} \text{ watts (or in log form} = -174 \text{ dBm)}$$

where

K = Boltzmann's constant = 1.381×10^{-23} W/Hz K
T = 290 K at room temperature
B = normalized bandwidth of 1 Hz

When the bandwidth is taken into account, the input noise level is calculated by:

$$\text{Input noise level} = 10 \; \log(KTB)$$

$$= 10 \; \log_{10}(B) - 174 \quad \text{(dBm)}$$

where B = bandwidth (Hz). For a 2 GHz bandwidth, from the above, there is -81 dBm noise at the receiver.

Table 1.8 is an example of a millimetre wave communication system link budget. Based on this table, the signal-to-noise ratio of the system can be calculated.

Table 1.9 is an example of a cascaded millimetre wave receiver, which includes a feeding network, a bandpass filter, a low-noise amplifier, a switch and channel selection filter, and an amplifier. The gain and noise figures for each component are provided, and the cumulative gain and noise figures are calculated.

Table 1.9 Components and their gain / noise figures

	Feeding network	Bandpass filter	LNA	Switch and channel selection filter	Amplifier
Gain (dB)	-5	-1	20	-5	30
Cumulative gain (dB)	-5	-6	14	9	39
Cumulative gain (real)	0.31	0.25	25	7.94	7943
Noise figure (dB)	5	1	3	5	10
Noise figure (linear)	3.16	1.26	2.00	3.16	10.00
Cumulative noise figure (linear)	3.16	3.98	7.94	8.03	9.16
Cumulative noise figure (dB)	5	6	9	9.04	9.62

A typical cascaded millimetre wave system is illustrated in Table 1.10. The transmit power is assumed to be 10 dBm, and the loss for the feeding network for the transmitting antenna is assumed to be 5 dB. 15 dBm of power should be achieved before the signal enters the feeding network. The transmitting antenna has a gain of 12 dBi, so the effective isotropic radiated power (e.i.r.p) increases to 22 dBm. During propagation, the signal undergoes free space loss and reflection loss, and so is reduced to -75 dBm. After the 12 dBi gain of the receiving antenna and the 5 dB loss of its feeding network, the signal increases to -68 dBm. Then the signal then passes through a filter with a -1 dB loss, a low-noise amplifier with a 20 dB gain (-43 dBm), a selection filter with a -5 dB loss (-48 dBm) and an amplifier with a 30 dB gain. Finally, the signal power is -18 dBm.

The input noise level in Table 1.10 is stated as -81 dBm, as the bandwidth is assumed to be 2 GHz. The noise then increases to -58 dBm due to a low-noise amplifier with a 3 dB noise figure, which is then reduced to -63 dBm due to the selection filter with a 5 dB loss. Therefore, the signal-to-noise ratio at the output of the selection filter is $(-54) - (-63) = 9$ dB. The power level is plotted in Figure 1.21.

Table 1.10 Spreadsheet for a cascade of millimetre wave circuits

	Feeding network (Tx)	Tx antenna	Free space loss	Reflection loss	Rx antenna	Feeding network (Rx)	RF bandpass filter	LNA	Channel selection filter	Amplifier	Output signal level (dBm)	Output noise level (dBm)
Input signal level (dBm)	15	10	22	−59.9842	−74.9842	−62.9842	−61.9842	−62.9842	−42.9842	−47.9842	−17.984	
Input noise level (dBm)						−80.9897	−80.9897	−80.9897	−57.9897	−62.9429		−32.37
Input SNR (dB)						18.0055	19.0055	18.0055	15.005503	14.95869	14.385	
Gain (dB)	−5	12	−81.9842	−15	12							

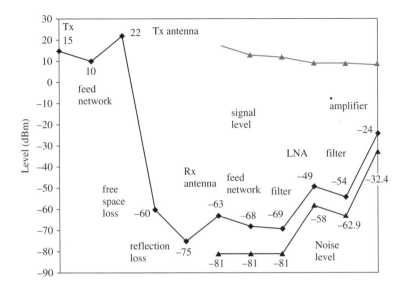

Figure 1.21   The 60 GHz link budget from Tx to Rx

1.8 Summary

This chapter explained the overall ideas and the importance of a gigabit wireless communication system using millimetre wave technology. A number of application scenarios are discussed. The international standards and regulations are compared, and the communication system concept is analysed. The role of antennas in the gigabit communication system is discussed. The characteristics of millimetre waves are addressed and a measured propagation result and channel performance are presented. The technical challenges for different antennas are investigated. Finally, an example of the link budget was provided, to show the performance of the system. Noise and its impact on link behaviour are also considered.

References

[1] Jeff Caruso, 'Copper 10 Gigabit Ethernet NICs Unveiled', *Network World*, January 2007
[2] Rick Merritt, 'New Tech Breaks into Network Specs War', *EE Times*, December 2006.
[3] http://www.ieee802.org/15/pub/TG3c.html
[4] http://www.wimedia.org/
[5] Kursat Kimyacioglu, 'WiMedia Next Gen UWB and 60 GHz Considerations', WiMedia Conference, March 2006.
[6] Su Khiong Yong 1 and Chia-Chin Chong, 'An Overview of Multigigabit Wireless through Millimetre Wave Technology: Potentials and Technical Challenges', *EURASIP Journal on Wireless Communications and Networking*, **2007**, 2007, Article ID 78907, 10 pp., DOI:10.1155/2007/78907.
[7] D. C. O'Brien, G. E. Faulkner, K. Jim and D. J. Edwards, 'Experimental Characterization of Integrated Optical Wireless Components', *IEEE, Photonics Technology Letters*, **18**(8), April 2006, 977–979.
[8] D. C. O'Brien, G. E. Faulkner, K. Jim, E. B Zyambo and D. J. Edwards, 'High-speed Integrated Transceivers for Optical Wireless', *IEEE, Communications Magazine*, **41**(3), March 2003, 58–62.
[9] Aki Tsukioka, 'JVC Develops Base Technologies for Next-Generation Optical Wireless Access System', *JCN Network*, 4 October 2005.

[10] A. M. Street, P. N. Stavrinou, D. C. O'Brien and D. J. Edwards, 'Indoor Optical Wireless Systems: A Review', *Optical and Quantum Electronics*, **29**(3), 1997, 349–378.

[11] D. C. O'Brien, E. B. Zyambo, G. Faulkner, D. J. Edwards, D. M. Holburn, R. J. Mears, R. J. Samsudin, V. M. Joyner, V. A. Lalithambika, M. Whitehead, P. Stavrinou, G. Parry, J. Bellon and M J. Sibley, 'High-Speed Optical Wireless Transceivers for In-building Optical Local Area Networks (LANs)', Conference on '*Optical Wireless Communications III*', 4124, paper 4124-16, SPIE, Boston, Massachusetts, 2000.

[12] Kao-Cheng Huang and Zhaocheng Wang, 'Millimetre-Wave Circular Polarized Beam-Steering Antenna Array for Gigabit Wireless communications', *IEEE Transactions on Antennas and Propagation*, **54**(2), Part 2, February 2006, 743–746.

[13] R. C. Qiu, H. Liu, X. Shen, 'Ultra-wideband for Multiple Access Communications', *IEEE Communications Magazine*, **43**(2), February 2005, 80–87.

[14] JVC Products, VIPSLAN OA-301, JVC Corporation Japan, http://www.jvc.co.jp/

[15] Arturas Medeisis, 'SE19 Drafting Group Meeting on MGWS at 60 GHz, ERO, Copenhagen, 26 March 2007', ERO SE19 Broadband Applications in Fixed Service, http://www.ero.dk/

[16] European Radiocommunications Committee (ERC), T/R 22-03E, 'Provisional Recommended Use of the Frequency Range GHz by Terrestrial Fixed and Mobile Systems', 1990, p. 3.

[17] CEPT, ERO, 'The European Table of Frequency Allocations,Locations and Utilisations Covering the Frequency Range 9 kHz to 275 GHz', Lisboa, January 2002; Dublin, 2003; Turkey, 2004; Copenhagen, 2004.

[18] ERC Recommendation 12-09, 'Radio Frequency Channel Arrangement for Fixed Service Systems Operating in the Band 57.0–59.0 GHz Which Do Not Require Frequency Planning', The Hague, 1998; revised Stockholm, October 2004.

[19] ECC Recommendation (05)02, 'Use of the 64–66 GHz Frequency Band for Fixed Services', June 2005.

[20] ETSI DTR/ERM-RM-049, 'Electromagnetic Compatibility and Radio Spectrum Matters (ERM); System Reference Document; Technical Characteristics of Multiple Gigabit Wireless Systems in the 60 GHz Range', March 2006.

[21] IEEE 802.15-15-06-0044-00-003c Document, '60 GHz Regulation in Germany', January 2006.

[22] Japan Regulations for Enforcement of the Radio Law 6-4-2 Specified Low Power Radio Station (11) 59–66 GHz Band.

[23] Ministry of Information Communication of Korea, 'Frequency Allocation Comment of 60 GHz Band', April 2006.

[24] FCC, 'Code of Federal Regulation, Title 47 Telecommunication', Chapter 1, Part 15.255, October 2004.

[25] Su Khiong Yong 1 and Chia-Chin Chong, 'An Overview of Multigigabit Wireless through Millimetre Wave Technology: Potentials and Technical Challenges', *EURASIP Journal on Wireless Communications and Networking*, **2007**, 2007, Article ID 78907, 10 pp.,DOI:10.1155/2007/78907

[26] Alireza Seyedi Philips, 'Proposed European Regulations', May 2006, IEEE 802.15-06-0247-00-003c.

[27] Spectrum Management Telecommunications, 'Radio Standard Specification-210, Issue 6, Low-Power License-Exempt Radio Communication Devices (All Frequency Bands): Category 1 Equipment', September 2005.

[28] FCC document, OMB 3060-1070, 'Allocations and Service Rules for the 71–76 GHz, 81–86 GHz, and 92–95 GHz Bands'.

[29] Jonathan Wells, 'Multigigabit Wireless Connectivity at 70, 80 and 90 GHz', *RF Design*, May 2006, 50–54.

[30] ERC Recommendation 12-09, 'Radio Frequency Channel Arrangement for Fixed Service Systems Operating in the Band 57.0–59.0 GHz Which Do Not Require Frequency Planning', The Hague, 1998; revised Stockholm, October 2004.

[31] ARIB STD-T69, 'Millimetre-Wave Video Transmission Equipment for Specified Low Power Radio Station', July 2004.

[32] ARIB STD-T74, 'Millimetre-Wave Data Transmission Equipment for Specified Low Power Radio Station (Ultra High Speed Wireless LAN System)', May 2001.

[33] http://www.ieee802.org/15/pub/TG3c.html

[34] IEEE 802.15-05-0353-07-003c, 'Working Group for Wireless Personal Area Networks (WPANs), TG3c System Requirements', January 2007.

[35] H. T. Friis, 'A Note on a Simple Transmission Formula', Proceedings of the IRE, **34**, 1946, 254–256.

[36] M. Marcus and B. Pattan, 'Millimetre Wave Propagation; Spectrum Management Implications', *IEEE Microwave Magazine*, **6**(2), June 2005, 54–62.

[37] David A. Sobel, '60 GHz Wireless System Design: Towards a 1Gb/s wireless link', Research Retreat at Berkeley Wireless Research Centre, June 2003.

[38] David A. Sobel and Robert W. Brodersen, '60GHz CMOS System Design: Challenges, Opportunities, and Next Steps', Research Retreat at Berkeley Wireless Research Centre, January 2003.

[39] M. K. Simon and M. S. Alouini, *'Digital Communication over Fading Channels'*, 2nd edition, Wiley-IEEE Press, New York, 2004.

[40] Katsuyoshi Sato, Takeshi Manabe, Toshio Ihara, Hiroshi Saito, Shigeru Ito, Tetsu Tanaka, Kazuyoshi Sugai, Norichika Ohmi, Yasushi Murakami, Masanori Shibayama, Yoshihiko Konishi and Tsuneto Kimura, 'Measurements of Reflection and Transmission Characteristics of Interior Structures of Office Building in the 60-GHz Band', *IEEE Transactions on Antennas and Propagation*, **45**(12), December 1997, 1783–1792.

[41] Chinh H. Doan, Sohrab Emami, David A. Sobel, Ali M. Niknejad and Robert W. Brodersen, 'Design Considerations for 60 GHz CMOS Radios', *IEEE Communications Magazine*, December 2004, 132–140.

[42] M. R. Williamson, G. E. Athanasiadou and A. R. Nix, 'Investigating the Effects of Antenna Directivity on Wireless Indoor Communication at 60 GHz', 8th IEEE International Symposium PIMRC, September 1997, pp. 635–639.

[43] R. C. Hansen, *'Phased Array Antennas'*, Wiley-Interscience, 19 January 1998.

[44] A. Abidi, 'Direct Conversion Radio Transceivers for Digital Communications', *IEEE JSSSC*, **30**(12), 1995, 1399–1410.

[45] F. Aschwanden, 'Direct Conversion – How to Make It Work in TV Tuners', *IEEE Transactions on Consumer Electronics*, **42**(3), August 1996, 729–751.

[46] J. Wenin, 'ICS for Digital Cellular Communication', European Solid State Circuits Conference, Ulm, Germany, 1994, pp. 1–10.

[47] Y. Zhao, C. Viereck, J. F. Frigon, R. G. Bosisio and K. Wu, 'Direct Quadrature Phase Shift Keying Modulator Using Sixport Technology', *Electronics Letters*, **41**(21), 2005, 1180–1181.

[48] R. G. Vaughan, N. Scott and D. White, 'The Theory of Bandpass Sampling', *IEEE Transactions on Signal Processing*, **39**(9), September 1991, 1973–1984.

[49] C. A. Balanis, *'Antenna Theory: Analysis and Design'*, 2nd edition, John Wiley & Sons, Inc., New York, 1997.

[50] K.-C. Huang and Z. Wang, 'Millimetre-Wave Circular Polarized Beam-Steering Antenna Array for Gigabit Wireless Communications', *IEEE Transactions on Antennas and Propagation*, **54**(2), Part 2, 2006, 743–746, DOI:10.1109/TAP.2005.863158.

[51] J.-Y. Park, Y. Wang and T. Itoh, 'A 60 GHz Integrated Antenna Array for High-Speed Digital Beamforming Applications', in Proceedings of IEEE MTT-S International Microwave Symposium Digest, Vol. 3, Philadelphia, Pennsylvania, June 2003, pp. 1677–1680.

[52] K. Huang and S. Koch, 'Circular Polarization Antenna', European Patent EP1564843.

[53] K. Huang and Z. Wang, 'Dielectric Rod Antenna and Method for Operating the Antenna', World Patent WO2006097145; European Patent EP1703590.

[54] IEEE 802.15-05-0493-27-003c, 'TG3c Selection Criteria', January 2007.

2

Critical Antenna Parameters

Based on the discussion in Chapter 1, the descriptions and requirements are developed, for the system components of a (nominally) 60 GHz free space point-to-point communications system. Also to be taken into consideration for this technology are user-defined constraints (such as size and bulk) and these will dictate or steer the direction of the design philosophy. There are five main constraints that will be explored for millimetre wave antenna design.

The first constraint is that the 60 GHz channel is lossy (due to oxygen absorption) but is otherwise benign. The excess loss at 60 GHz is approximately 15 dB/km and it is therefore desirable to identify means to overcome oxygen absorption and ensure that a sufficient margin exists to overcome other losses, such as rain-induced fading. Here compensation can be acheived by increasing the transmitter or receiver antenna gain. For example, a directional antenna gain can be employed to substitute for raw transmitter power and receiver noise. Thus there is the prospect of system optimisation by trading off the requirements in these different areas, and this aspect will be considered in Section 2.1.

The second constraint for systems is a strong multipath effect in an indoor environment. In other words, the line-of-sight signal and the reflected signal will arrive at the receiver via different paths. When the path difference is $n \times \lambda/2 (n = 1, 3, \ldots)$, there is a destructive interference between signals and this causes a notch in the frequency spectrum. For example, if the path difference is 2.5 mm, there will be a notch at 60 GHz and such a notch can cause an unstable wireless link with only slight physical displacement of the terminals or scatterers; this will reduce the quality of the communications link. To minimise the multipath effect, a narrow-beam antenna is therefore preferred. A discussion of the optimisation of the beamwidth is given in Section 2.2.

The third constraint is the space limit for portable devices such as handsets. It is essential to know how much gain can be achieved for an antenna in such a restricted space limit. This limit will be discussed in Section 2.3. For a predominantly line-of-sight wireless link, circular polarisation is useful to filter out the first reflected (multipath) signal. Additionally, the wireless communication data rate (the capacity of the link) can be increased by using multipolarisation in a multitransmitter–multireceiver system. Each polarization state can deliver different information channels and thus the data rate will be increased two to three times. Conversely, if the major concern is the robustness of the link, the question of frequency re-use by polarisation

Millimetre Wave Antennas for Gigabit Wireless Communications Kao-Cheng Huang and David J. Edwards
© 2008 John Wiley & Sons, Ltd

can be employed to support multiple copies of the information channel. The polarization will be discussed in Section 2.4.

Fourthly, there is the problem of noise and interference. The reliability of the communications link is defined by the signal-to-noise ratio. In the general sense, any undesired power appearing in the communications channel is noise and degrades the performance of the link. The sources of noise are various and can be due to environmental radiators (either passive or active) and that generated in the transceivers. This last component can only be controlled by good design practice and is not the main consideration of this work. The environmental component can, however, be mitigated by the antenna performance and this will be a consideration in Chapter 7.

Finally, the wireless link can be interrupted by a blocking object (such as a human body) which introduces additional shadowing loss. To avoid this shadowing loss, a beam-steering function or multibeam antennas can be considered to cope with this loss. The approaches using these types of antennas will be discussed in a later chapter.

2.1 Path Loss and Antenna Directivity

For a wireless LAN, it is generally assumed that the signal arriving at the receiver consists of many copies of the information-carrying signal, which have been generated by scattering and other processes by the environment. Each path will have a specific delay, and arrival times will vary according to the dimensions of the environment.

For a specific path, as shown in Figure 2.1, the delay profile of the channel is determined by the delay time, path gain and phase of each path. In order to reduce the multipath effect, it is usual to receive each copy or path and time shift (and/or phase shift), in order to maximise the received power and reduce the distortion of the signal; it is usually only necessary to reduce the effects of the major paths that have the most power (usually no more than the first four or five).

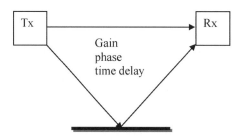

Figure 2.1   Path channel model

In a gigabit wireless system, the channel model is often assumed to be quasi-optical and indeed line-of-sight (LOS); therefore the major power is in the direct path in this case. A typical indoor measurement of 60 GHz propagation is shown in Figure 2.2. It can be seen that there can be significant delayed components due to reflections and if there is no direct line-of-sight then comparable power may be distributed over many reflections. In addition, Figure 2.3 plots the free space loss as a function of distance at 37 GHz for similar measurements. These two figures show that a strong path loss exists in the range of the millimetre wave spectrum.

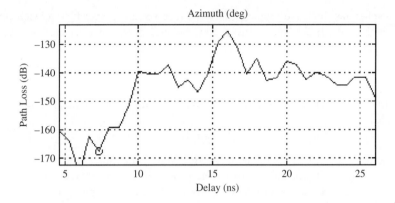

Figure 2.2 Path loss and delay time for the 60 GHz signal [1]

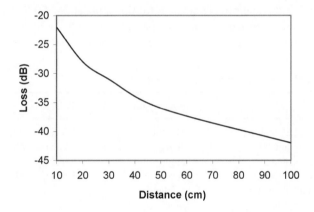

Figure 2.3 Path loss as a function of Tx–Rx separation at 37 GHz [2]

To combat these effects, it is necessary to focus or direct the radiated power from antennas in a given direction. The power flux density in this direction will be greater than if it were an omnidirectional antenna transmitting the same power (the power presented at the antenna input terminals) and the ratio between these values (i.e. the degree to which the antennas enhances the power flux density relative to an isotropic radiator) is called the antenna gain [3]. The degree to which the power is confined is called the directivity (or how directional the antenna is). These two quantities are closely related by the radiation efficiency of the antenna and can be expressed as follows:

$$\text{Gain} = \text{efficiency} \times \text{directivity}$$

A typical antenna is able to couple energy to and from free space with an efficiency of approximately 65 %.

The usual approach to establish the power received by an antenna is to consider an iso-tropic radiator transmitting power P_T; so that this power is distributed over the surface of an expanding sphere as the wave propagates. At the receiver, the power flux density (power per unit area) is then $P_T/4\pi R^2$. The received power is then determined by the effective capture area of the receive antenna, so that the power received is then:

$$A_E P_T/4\pi R^2$$

This effective capture area can in turn be related to the gain of the antenna [3], so the gain can be written as:

$$G = 4\pi A_E/\lambda^2$$

As can be seen, the directivity of an antenna is given by the ratio of the maximum radiation intensity (power per unit solid angle) to the average radiation intensity (averaged over a sphere). The directivity of any source, other than an isotrope, is always greater than unity. The maximum gain of an antenna is simply defined as the product of the directivity and its radiation efficiency. If the efficiency is not 100 %, the gain is less than the directivity. When the reference is a lossless isotropic antenna, the gain is expressed in dBi (decibels relative to an isotrope).

To aid in designing the appropriate antenna for the application, Table 2.1 lists the major technologies for millimetre wave antennas and provides a comparison of the features for the different types of antenna. More details can be found in the following chapters.

Table 2.1 Comparison of different types of antennas

	Power gain	Polarisation
Printed antenna	Medium	Linear/circular
Horn antenna	High	Linear
Lens antenna	High	Linear
Rod antenna	High	Linear/circular
Helical antenna	Medium	Circular
Multidipole	Medium	Linear/circular
Dipole	Low	Linear
Slot antenna	Low	Linear/circular

In gigabit wireless communications, low profile design is attractive due to ease of fabrication, and such a design has the potential to be built at low cost. Furthermore, the structures can be lighter than reflector antennas of similar performance and also easier to install.

Several configurations have been proposed for this type of application in recent years that produce high directivity at broadside (the direction perpendicular to the antenna's length) [3, 4]. In the following chapters an historical overview of such configurations will be presented and then illustrations given of the fundamental principles of operation, fabrication and testing. Also discussed will be some new ideas that have emerged in the past few years with the use of electromagnetic bandgaps (EBGs), metamaterials and metasurfaces to extend the performance of previous designs.

As long ago as 1956, the first high-directivity "flat" antenna was designed to produce the high directivity at broadside excited by a single source, as shown in Figure 2.4 [4]. It employs a partially reflective surface (PRS) located approximately a quarter-wavelength above a ground plane. Such a structure builds a Fabry–Perot cavity (FPC) and successive reflections of the trapped energy escape as a coherent summation along a defined direction, thus producing a sharp beam. Subsequent research articles were published in 1985 [5] and in 1988 [6], where the reflective surface was replaced with a dense, quarter-wavelength, dielectric, still over a half-wavelength cavity. In 1988, the idea that the excitation of a leaky-wave contributes to the high directivity was developed using a single dielectric layer for the first time. Later in 2001, the concept introduced by Von Trentini was generalised to other geometries and actual designs were made [7]. Utilising the same concepts, the partially reflective surface layer proposed initially by Von Trentini was generalized to more exotic periodic structures (metasurfaces) [7]. The resonant frequency could be controlled and thus the operating bandwidth, which was in line with modern communication systems requirements.

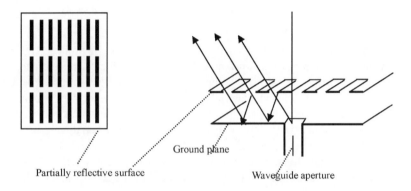

Partially reflective surface

Ground plane

Waveguide aperture

Figure 2.4 Schematic diagram of the antenna and geometry of dipole partially reflective surfaces [4, 8]

In addition to partially reflective surfaces, studies have shown that resonant defects in an electromagnetic bandgap material could be used to produce high directivity outside the crystal [9–14]. In practice, in these studies the reflective superstrate was replaced by a single or multiple layers of electromagnetic bandgap material, but still over a resonating cavity.

The antennas described above could be excited by a single source located inside the cavity, such as a coaxial probe, a microstrip patch, a slot in the ground plane or by a waveguide horn. Some examples in the context of the fabrication of prototypes will be shown in the next few chapters, illustrating the performance of such antennas in terms of bandwidth, aperture efficiency, etc.

The above optical concepts have been applied to design a Fabry–Perot cavity (FPC) that encapsulates a dual polarised array with sparse elements [15, 16] (see Figure 2.5). The unusually large distance between array elements allows the design of the beamforming network on the same plane of the array. This approach also supports dual polarisation.

A further advance in research work has shown that a metamaterial slab made of parallel wires excited by a single source could produce a high directivity [17, 18]. As shown also by other researchers, a material made of wires exhibits the electromagnetic behaviour of a plasma, with a plasma cut-off frequency that depends on the radius of the wires and on the period of the structure. A similar idea was reported in Reference [19], which studied and realised a metamaterial made of wire grids that produce high directivity. A detailed explanation of the phenomena was presented in Reference [20], which also provided design criteria for directivity and bandwidth.

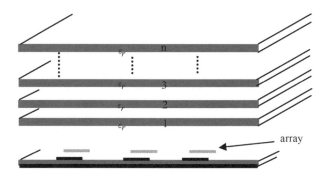

Figure 2.5 Cut view of a sparse array in a Fabry–Perot cavity

It was shown that a highly directive beam in this class of antenna is also produced by an excited leaky-wave with a small attenuation constant and large phase velocity. In Reference [21], the leaky-wave model of Reference [20] was compared with a ray-optic description, as illustrated in Figure 2.6. A larger class of metamaterials for directivity enhancement was analysed in Reference [22], where low and high permittivity and permeability materials and the concept of low and high impedance materials were analysed, to produce enhanced directivity for a given direction. A brief comparison will now be made of this class of antenna with those previously described with the partially reflective surface, as was reported in Reference [23], where the figure of merit was introduced and taken as the product of the directivity and the bandwidth.

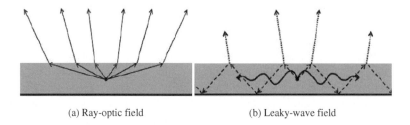

(a) Ray-optic field   (b) Leaky-wave field

Figure 2.6 Radiation mechanisms for the directive broadside radiation.(a) Ray-optic model showing the refractive lensing effect at the top interface. (b) Leaky-wave model showing a propagating leaky mode that is excited by the line source. (Reproduced by permission of © 2006 John Wiley & Sons, Inc. [21])

For an antenna with a 20° half-power beamwidth (symmetric), the directivity can be calculated approximately to be:

$$D = \frac{40\,000}{\theta_{\text{HPBW}}\phi_{\text{HPBW}}} = \frac{40\,000}{20 \times 20} = 100 \text{ or } 20\,\text{dBi}$$

To have such high directivity and gain in a small size, one of the solutions is to combine both a planar antenna and a three-dimensional antenna. Numbers from various research outcomes are summarised in Table 2.2.

Table 2.2 Combination of two- and three-dimensional antennas to increase antenna gain

3D antenna 2D antenna	Rod	Reflector	Lens	Horn
Patch	Ref. [24]	Ref. [25]	Ref. [26]	Ref.[28]
Slot	–	Ref. [27]	Figure 2.8	Figure 2.7
Yagi	–	–	–	Figure 2.9

Some examples from Table 2.2 will now be discussed in more detail. A slot antenna can be combined with a horn antenna as illustrated in Figure 2.7. Similarly, a patch-fed horn antenna has been experimentally investigated at microwave and millimetre wave frequencies [28]. The results indicated that for a 70° flare-angle horn, horn apertures from 1.0 λ-square to 1.5 λ-square, with dipole positions between 0.36 and 0.55 λ, yield good radiation patterns with a gain of 10–13 dB at 60 GHz, and a cross polarisation level lower than −20 dB on bore sight. It also found that the impedance measurements can be reliably used for two-dimensional horn arrays, but the radiation patterns differ because of the Floquet modes [29] associated with the array environment. The integrated horn antenna is a high-efficiency antenna suitable for applications in millimetre wave imaging systems, remote sensing and radio astronomy.

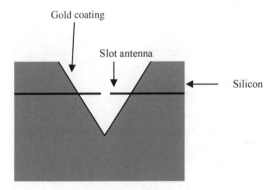

Figure 2.7 The complete millimetre wave antenna structure

The antenna gain can be increased by integrating slot antennas with a dielectric lens. One example is shown in Figure 2.8. The lens is made out of low-cost low-permittivity Rexolite material. The single-beam lens achieves a gain of 24 dBi at 30 GHz and a front-to-back ratio of 30 dB. An axial ratio of 0.5 dB is maintained within the main lobe [30]. The measured impedance bandwidth is 12.5 % within a standing-wave ratio (SWR) of 1.8 : 1. The single-beam antenna is well suited for broadband wireless point-to-point links.

In Figure 2.8, a lens, which is fed by multiple slots, can radiate multiple beams with a minimum 3 dB overlapping level among adjacent beams. The coverage of the lens antenna system has been optimised through the utilisation of a number of slot arrangements, leading to broad scan coverage. The multiple-beam lens antenna is suitable for an indoor wireless access point or as a switched beam smart antenna in portable devices. More details about lens antennas can be found in Chapter 6.

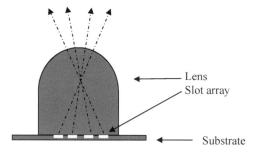

Figure 2.8   Multiple-beam launching through a substrate lens antenna

It is also possible to excite a circular horn antenna with a quasi-Yagi antenna, as illustrated in Figure 2.9 [31]. Single-mode operation was achieved by placing the circular waveguide transition in the horn, which suppresses the potential excitation of higher-order modes. A typical aperture efficiency of 60 % at 60 GHz for a single-mode circular horn antenna was achieved due to the high radiation efficiency of the quasi-Yagi antenna. The measured antenna gain and radiation patterns of the longer horn, correspond to optimum horn characteristics with a waveguide input. A wider bandwidth can be achieved by realising the transition in the waveguide, which feeds the horn.

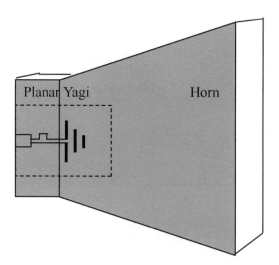

Figure 2.9   Cross-section of a circular horn antenna with a quasi-Yagi antenna inside. (Reproduced by permission of © 2001 IEEE [31])

The integration of a quasi-Yagi antenna with a horn makes this antenna a symmetric two-port device regardless of the angle of reception, which can be realised in balanced receivers and transmitters. The edge diffraction from the incoming horn aperture is reduced, which can be of use in corrugated horns. The single-mode operation of the antenna allows the integration of a polariser directly at the aperture.

When implementing a single-element antenna (such as a dielectric lens or a slot), it is possible to use two-antenna systems to achieve up to a 75 % reduction of the lens material whilst maintaining about the same length and on-axis characteristics, as shown in Figure 2.10 [32]. The lens-fed reflector provides higher overall efficiency than the two-lens system. This makes the lens-fed reflector attractive for single-beam applications. In these two-antenna systems, a limited scan capability with multiple beams cross-coupled at the 3 dB level is possible, which can lead to lower alignment requirements between a receiver and a transmitter for line-of-sight broadband wireless links.

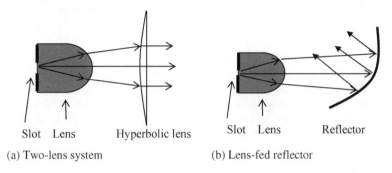

Slot   Lens   Hyperbolic lens   Slot   Lens   Reflector

(a) Two-lens system   (b) Lens-fed reflector

Figure 2.10   Schematic diagram of the two-antenna systems

2.2 Antenna Beamwidth

The radiation pattern of an antenna is essentially the Fourier transform (linear space to the angle) of its aperture illumination function. In a radiation pattern cut containing the mainlobe direction, the angle between the two directions, where the radiation intensity is one-half the maximum value, is called the half-power beamwidth (see Figure 2.11).

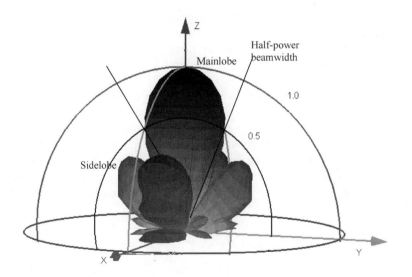

Figure 2.11   Beamwidth in a normalised power pattern (the radial scale is logarithmic)

The beamwidth of an antenna is a measure of the directivity of an antenna and is usually defined by the angles where the pattern drops to one-half of its peak value; they are also known as the 3 dB points. For a circular aperture antenna of diameter D, if the antenna is uniformly excited, this beam width is about $70 \times$ wavelength$/D$ (the exact beamwidth actually depends on the aperture illumination function). The next lobe in the pattern, usually called the first sidelobe, will be about 1/20 (13 dBs less) of the value of the main lobe, and any further out sidelobes will have an even lower value. The rate of decay of these sidelobes is an important parameter in many antenna applications and is used in many international standards as a defining parameter of antenna performance. In general, the maximum gain can be approximated by the following formula:

$$G = \frac{27\,000}{BW_h \times BW_v}$$

where G is the power gain (linear) of the antenna, BW_h is the horizontal beamwidth of the antenna and BW_v is the vertical beamwidth of the antenna [1].

As an example, consider an antenna that has a vertical beam width of 27° and a horizontal beam width of 10°; it will have a power gain of 100 (linear) or 20 dB. It would also have a vertical dimension of about 2 wavelengths and a horizontal dimension of about 5 wavelengths if the antenna is uniformly excited.

The total received signal is normally expressed as a closed-form expression, known as the Friis equation [1]:

$$P_{Rx} = P_{Tx} \frac{G_{Tx} G_{Rx} \lambda^2}{16 \pi^2 d^2 L}$$

where P_{Rx} is the received power, P_{Tx} is the transmitted power, G_{Tx} and G_{Rx} are the antenna transmit gain and receive gain, respectively, and λ, d and L are the wavelength, separation and other losses, respectively. The allocations given to each of these components constitute what is generally called the link budget.

There are four types of antenna configurations in a communications system:

1. Tx: omnidirectional antenna versus Rx: omnidirectional antenna
2. Tx: omnidirectional antenna versus Rx: directional antenna
3. Tx: directional antenna versus Rx: omnidirectional antenna
4. Tx: directional antenna versus Rx: directional antenna

Omnidirectional antennas have signals radiating in all directions and are useful when a multipath is needed for communication purposes. A directional antenna has a narrow beam in the desired direction and receives less well in the undesired direction. This is useful when a multipath is not required. As a directional antenna has small coverage, it may be necessary to incorporate it with a beam-steering function to provide wider coverage. However, the narrower the antenna beamwidth, the more complex the beam-tracking function would need to be. Hence it is necessary to consider the balance between the complexities of a beamforming antenna and a tracking function.

2.3 Maximum Possible Gain-to-Q

In millimetre wave applications it is desirable to maximize antenna gain and bandwidth (i.e. to minimize the Q for a lossless high gain antenna) simultaneously. Therefore the optimisation of the ratio of the gain-to-Q is important in antenna design. It is clear that the optimisation of the ratio of the gain-to-Q will yield a greater minimised Q than the minimum possible Q discussed previously, since it demands the gain to be maximised at the same time.

The quality factor Q of an antenna is an important overall parameter specifying the antenna performance and the inherent physical limitations of antenna size on the gain. In particular, a high value of Q means that large amounts of reactive energy are stored in the near zone field. This in turn implies large currents, high ohmic losses, narrow bandwidth and a large frequency sensitivity. Knowledge of the antenna Q leads to a reasonably definite assessment of antenna performance because of its clear physical implications.

The first general study was published by Chu [33], who derived theoretical values of Q for an ideal antenna enclosed in an imaginary sphere. The Q of an electrical network at resonant frequency ω can be defined as:

$$Q = \frac{\omega W_1}{P_1} \tag{2.1}$$

where W_1 is the time-average energy stored in the network, and P_1 is the power dissipated in the network. When the network is not resonant, the time-average magnetic energy stored in the network is not the same as the time-average electric energy. The input impedance is proportional to $P + 2j\omega(W_m - W_e)$, where W_m and W_e are the time-average magnetic and electric energy respectively, stored in the network.

To make the input impedance resistive, some additional energy storage must be added so that the net reactive energy vanishes. If it is agreed that the network is always to be operated with an additional ideal lossless reactive element, so that the input impedance is purely real, then Q of the resultant network is defined as:

$$Q = \frac{2\omega W_2}{P} \tag{2.2}$$

where W_2 is the larger of W_m or W_e and P is the power dissipation in the original untuned network.

Therefore, the antenna is normally tuned to resonance by the addition of a reactive element. If the added reactive element does not dissipate any energy, Q of the resultant system is then given by Equation (2.2), where P is now the total radiated power and W_2 is the larger of W_m or W_e, these now being interpreted as the time-averaged magnetic and electric energy stored in the near zone field around the antenna. Consequently, Equation (2.2) can now be considered to be the upper bound on Q of an antenna system that can be tuned to resonance by the addition of a single reactive element. Any loss in the tuning element would reduce Q to a value below that given by Equation (2.2). From now on, the parameter defined by Equation (2.2) will be referred to as the antenna Q, even though in the usual network sense it is the Q that results only when the antenna is tuned to resonance by an ideal reactive element.

Looking at the field in the vicinity of a high-gain, small-size radiator, extremely large field intensities can be found. In a physical antenna, this would result in prohibitive heat loss. In an ideal lossless antenna, this would result in large energy densities, and would thus yield a high Q for an antenna. It is possible that the required current distribution cannot be obtained in practice, since it is determined by the solution to a boundary value problem. Assuming that the antenna geometry and excitation are arbitrary, the new upper limits can be derived for a directional antenna in this case.

The field external to a sphere containing all sources can be expanded in terms of a spherical wave coordinate system (Figure 2.12) [34]:

$$\psi_n = h_n^{(2)}(kr) f_n(\theta, \phi)$$

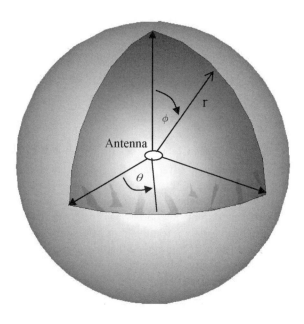

Figure 2.12   Spherical coordinate system

where n is an integer and $h_n^{(2)}$ is the spherical Hankel function (see [34]). With the miniaturisation of electronic devices, reductions in antenna size and profile are being continually demanded. A question that is frequently asked is how small an antenna can be made, whilst at the same time maintaining good performance (i.e. the highest gain and bandwidth at the same time). For a small antenna with $ka \ll 1$, the optimized quantities can be approximated as [35]:

$$\max \left. \frac{G}{Q} \right|_{dir} \approx \frac{6(ka)^3}{2(ka)^2 + 1} \tag{2.3}$$

Here k is the wave number in radians/mm and a is the free space wavelength in mm. It should be noted that an infinitesimally small dipole itself has an extremely narrow bandwidth, since its real Q would be much higher than the minimum possible Q (mathematically it should be infinity). The above relationships are the best overall performances a small antenna can achieve and they can be used to determine the smallest possible antenna size once the required antenna bandwidth is given. To illustrate the best antenna performances when Q is large, the maximum possible ratio of gain-to-Q, and the maximized gain is depicted in Figure 2.13. It is seen from the plot of $\max G/Q|_{dir}$ that they are all monotonically increasing functions of ka.

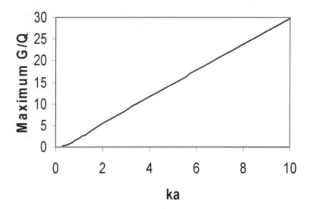

Figure 2.13 Maximum ratio of a gain-to-Q for a directional antenna

Figure 2.14 shows the plots of maximum gain of an antenna. Again these are monotonically increasing functions of ka. In general, the curve in Figure 2.14 is applicable to all antennas. They can be used to determine the best overall performances once the maximum antenna size is given, or to determine the smallest possible required antenna size to get the best overall performances, as already discussed for small antennas. It is well known that there is no mathematical limit to the gain that can be obtained from currents confined to an arbitrary small volume. However, a small-sized antenna with extremely high gain will produce high field intensity in the vicinity of the antenna, which results in high heat loss or high stored energy. By artificially truncating the spherical wave function expansions of the fields to the order N, the maximum gain obtainable is shown as [36]:

$$G_{\mathrm{max}} = N \times (N+2) \tag{2.4}$$

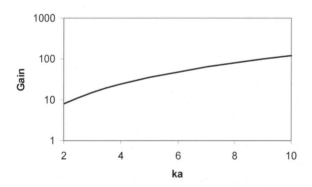

Figure 2.14 Maximised achievable gain. (Reproduced by permission of © 2003 IEEE [35])

Although Harrington obtained this result by considering a linearly polarized source, it can be easily proved that this result generally holds for an arbitrary current source. Hence as N increases (equivalently the antenna complexity increases) the maximum gain increases. Since the magnitude of the spherical Hankel function decreases very slowly for $n < ka$ and very rapidly for $n > ka$, the approximate transition point $ka = n$ can be considered to be the point of gradual cut-off [37].

The normal gain is often introduced and defined by letting $N = ka$ [33, 36, 38], so $G_{norm} = ka \times (ka+2)$. Any antenna having a larger gain than the normal gain has been called a supergain antenna. It is believed that using a supergain antenna will result in a high Q and is therefore not very practical, and so the normal gain is the maximum gain achievable without incurring high Q. The plot of G_{norm} is shown in Figure 2.14. It is clear that the definition of the normal gain or supergain is ambiguous since the cut-off point $n = ka$ is an approximate transition point and in addition, Q is not clearly specified.

Assuming there is an infinite ground plane and the sphere has a radius which can enclose the antenna, and $a = 5$ mm, the free space wavelength at 60 GHz can be calculated and the wave number that allows ka to be evaluated:

$$\lambda = \frac{c}{f} = \frac{3 \times 10^8 \times 10^3}{60 \times 10^9} = 5 \text{ mm}$$

$$k = \frac{2\pi}{\lambda} = 1.26 \text{ radians/mm}$$

Therefore:

$$ka = 1.26 \times 5 = 6.3$$

From Equations (2.3) and (2.4), the maximum gain-to-Q factor and maximum gain is given as:

$$\left. \max \frac{G}{Q} \right|_{dir} \approx \frac{6(6.3)^3}{2(6.3)^2 + 1} = 18.7$$

$$\text{Maximum gain} = 6.3(6.3 + 2) = 52.3 = 17 \text{ dB}$$

2.4 Antenna Polarisation

The subject of antenna polarisation has generated much published material over the years. The precise definition can be complex, and radiating and receiving structures respond varyingly, both in frequency and the angle to incident and transmitted waves. Here the discussion shall be confined to a simple treatment, and the reader is directed to texts that deal with the topic in much greater depth [1]. Only "far-field" radiation will be considered (since the wavelength is small compared with the dimensions of the radiators) and for the illustive cases presented here plane wave propagation will also be assumed.

In free space, the energy radiated by any antenna is carried by a transverse electromagnetic wave, that is comprised of an electric and a magnetic field. These fields are orthogonal to each another and also orthogonal to the direction of propagation. The electric field of the electromagnetic wave is used to define the polarisation plane of the wave, and therefore describes the polarisation state of the antenna.

For describing antenna polarisation properties the "Ludwig definition 3" is commonly used [37]. In this definition, reference and cross polarisations are defined as the measurement obtained when antenna patterns are taken in the usual manner, as illustrated in Figure 2.15.

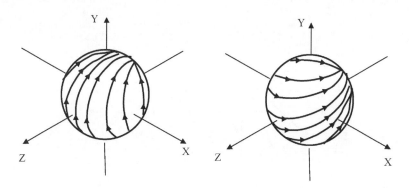

Figure 2.15 (a) Direction of the reference polarisation. (b) Direction of the cross polarisation (Ludwig 3)

This wave is said to be linearly polarised, i.e. the electric field vector is confined to a single plane. Two independent linearly polarised waves at the same frequency can therefore exist and propagate along the same path. This feature has been used for many decades in free space links, which utilise frequency re-use in order to double the capacity of a link for a given bandwidth. In this case, each polarisation carries different information and is transmitted and received independently. Where the relative angular orientation of the transmitter and receiver is not defined, using two linear polarisations becomes a problem as the alignment of the receiver with the transmitter is essential. Systems have been deployed in which dynamic control of the receiver is used with the incident linear polarisation, but coverage of these is beyond the scope of this book.

When the two polarisations carry the same information and the two components possess a specific phase relationship with each other, a form of wave can be constructed in which the electric field vector rotates as the wave propagates. If the relative phase of the two components is fixed at $\pm 90°$ and the amplitudes of the components are equal, the electric vector describes a circle as the wave propagates. Such a wave is said to be circularly polarised. The sense or handedness of the circular polarisation depends on the sense of the phase shift. In general, the two linear components of the propagating wave can have an arbitrary (though constant) phase relationship and also different amplitudes. Such waves are said to be elliptically polarised, as shown in Figure 2.16.

The majority of electromagnetic waves in real systems are elliptically polarised. In this case, the total electric field of the wave can be decomposed into two linear components, which are orthogonal to each other, and each of these components has a different magnitude and phase. At any fixed point along the direction of propagation, the electric field vector will trace out an ellipse as a function of time. This concept is shown in Figure 2.17, where, at any instant in time, E_x is the component of the electric field in the x direction and E_y is the component of the electric field in the y direction. The total electric field E is the vector sum of E_x plus E_y. The projection along the line of propagation is shown in Figure 2.17.

Therefore from the above discussion, there are two special cases of elliptical polarisation, which are linear polarisation and circular polarisation. The term used to describe the relationship between the magnitudes of the two linearly polarised electric field components in a circularly polarised wave is the axial ratio (AR). In a pure circularly polarised wave, both electric field components have equal magnitude and the AR is 1 or 0 dB (10 log [AR]). Thus in

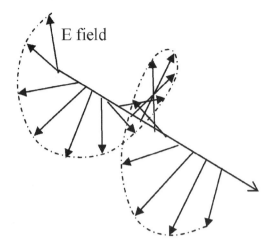

Figure 2.16   Propagation of elliptical polarisation

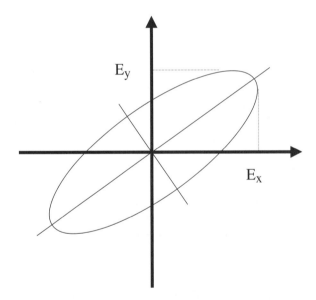

Figure 2.17   The projection of a polarisation ellipse

a pure linearly polarized wave the axial ratio is ∞. In this case, the polarisation ellipse traced by the wave is a circle.

It is difficult to make low cross polarisation circular sources that operate over a large bandwidth. Thus cross polarisation in the transmit antenna can be major source of error in antenna gain. To deliver maximum power between a transmitter and a receive antenna, both antennas must have the same angular orientation, the same polarisation sense and the same axial ratio. When the antennas are not aligned or do not have the same polarisation, there will be a reduction in energy or power transfer between the two antennas. This reduction in power transfer will reduce the overall signal level, system efficiency and performance. The polarisation loss can affect the link budget in a communications system.

It is reported that circular polarisation can reduce the power of the reflected path significantly in the millimetre wave LOS link [37, 39]. Owing to the boundary conditions on the electric field, the in-plane and normal components of the electric field suffer a differential phase shift of $180°$ on reflection. This causes the sense of the circular polarisation to be changed at each surface reflection. Thus for an odd number of reflections, the reflected wave attains a polarisation state orthogonal to the incident wave. When this occurs a left-hand circularly polarised wave would become a right-hand circularly polarised wave and vice versa. However, the direction of circular polarisation remains the same when there is an even number of reflections, and the power is only reduced due to reflection loss.

The circular polarisation maintains an advantage in some user scenarios. For example, if a user holds the terminal at an arbitrary tilt angle to the transmit signal, there would be a degradation of the signal strength in the case of linearly polarisation signals. However, such degradation is not present in the case of circular polarisation for a direct line-of-sight and arbitrary terminal tilt angle. In addition, the terminal will receive fewer multipaths (for a single polarisation), and circular polarisation also offers the possibility of frequency reuse, albeit with the complication of cross polar interference due to multipath reflections. Clearly the magnitude of the multipaths depends on the reflection coefficients of the materials in the environment; and on the material properties of the reflecting objects.

Conventional short-range systems normally use linearly polarised antennas to reduce cost. When the transmit and receive antennas are both linearly polarised, the physical antenna misalignment will result in a polarisation mismatch loss, which can be determined using:

$$\text{Polarisation mismatch loss (dB)} = 10 \log(\cos \theta) \quad (2.5)$$

where θ is the angular misalignment or tilt angle between the two antennas. Polarisation efficiency can be written as:

$$\text{Polarisation efficiency} = 20 \log \left(1 \pm \frac{A_1 - 1}{A_1 + 1} \frac{A_2 - 1}{A_2 + 1} \right)$$

where A is the axial ratio and the subscripts the antenna number. Figure 2.18 illustrates some typical mismatch loss values for various misalignment angles.

In the circumstance where the transmitting antenna in a wireless link is circularly polarised and the receiving antenna is linearly polarised, it is generally assumed that a 3 dB system loss will result because of the polarisation difference between the two antennas. In reality, the polarisation mismatch loss between these two antennas will only be 3 dB when the circularly polarised antenna has an axial ratio of 0 dB. The actual mismatch loss between a circularly polarised antenna and a linearly polarised antenna will vary depending upon the axial ratio of the (nominally) circularly polarised antenna.

When the axial ratio of the circularly polarised antenna is greater than 0 dB (i.e. it is in fact elliptically polarised), this will dictate that one of the two linearly polarised axes will generate a linearly polarised signal more effectively than the other component. When a linearly polarised receiver is aligned with the polarisation ellipse's major axis, the polarisation mismatch loss will be less than 3 dB. When a linearly polarised wave is aligned with the polarisation ellipse's linear

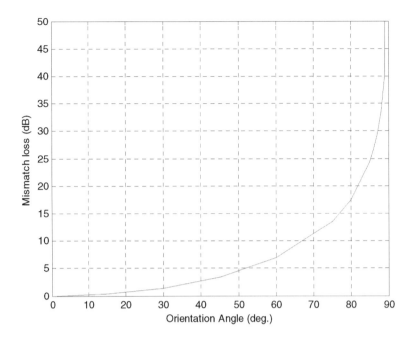

Figure 2.18   Polarisation mismatch between two linearly polarised waves as a function of angular orientation θ

minor axis, the polarisation mismatch loss will be greater than 3 dB. Figure 2.19 illustrates the minimum and maximum polarisation mismatch loss potential between an elliptically polarised antenna and a linearly polarised antenna as a function of the axial ratio. Minimum polarisation loss occurs when the major axis of the polarisation ellipse of the transmitter (receiver), is aligned with the plane of the linearly polarised wave of the receiver (transmitter). Maximum polarisation loss occurs when the weakest linear field component of the circularly polarised wave is aligned with the linearly polarised wave.

An additional issue to consider with circularly polarised antennas is that their axial ratio will vary with the observation angle [37, 39]. Most manufacturers specify the axial ratio at the antenna bore sight, or as a maximum value over a range of angles. This range of angles is generally chosen to represent the main beam of the antenna. In order to measure the axial ratio, antenna manufacturers measure the antenna radiation pattern with a spinning linearly polarised source. As the source antenna spins, the difference in amplitude between the two linearly polarised wave components radiated or received by the antenna is evident. The resulting radiation pattern will describe the antenna's axial ratio characteristics for all observation angles.

A typical axial ratio pattern for a circularly polarised antenna is presented in Figure 2.20. From the antenna radiation pattern, it can be seen that the axial ratio at the bore sight is about 0.9, while at an angle of $+60°$ off-bore sight, it dips to about 0.04. As the axial ratio varies with the observation angle, the polarisation mismatch loss between a circularly polarised antenna and a linearly polarised antenna will vary with the observation angle as well.

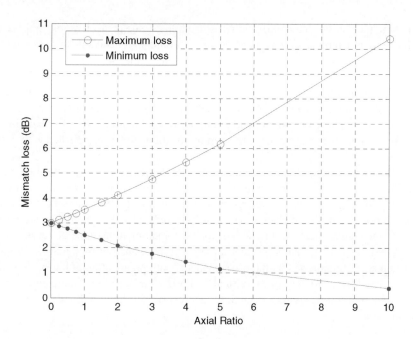

Figure 2.19 Polarisation mismatch between a linearly and a circularly polarised wave as a function of the circularly polarised wave's axial ratio

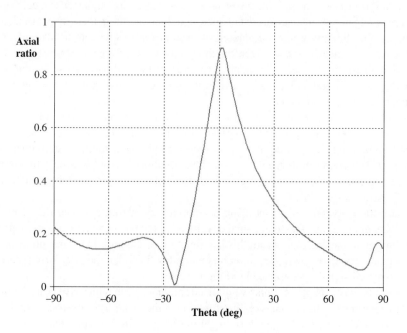

Figure 2.20 Typical axial ratio pattern for a helix antenna

However in most cases, the polarisation mismatch loss issue is much more complex. Based upon this discussion of polarisation mismatch loss, it would be conceivable that communication in a wireless system is near impossible when, for instance, the antenna of a mobile device is orthogonal to the antenna of an access point.

Obviously, this is unlikely to happen in the real world. In any mobile handset communications link, the signal between the handset antenna and the base station antenna is generally comprised of a direct line-of-sight (LOS) signal and a number of multipath signals. In many instances, the LOS signal is not present and the entire communications link is established with multipath signals.

Multipath signals arrive at the antennas of mobile devices via the reflection of the direct signal off nearby and distant objects or surfaces. If the reflecting objects are oriented such that they are not aligned with the polarisation of the incident wave, the reflected wave will experience a polarisation state change. The resultant, or total signal, available to the receiver at the end of the communications link will be the vector summation of the direct signal and all of the multipath signals. In many instances, there will be a number of signals arriving at the receive site that are not aligned with the assumed standard polarisation of the system antenna. As the receive antenna rotates from vertical to horizontal, it simply intercepts or receives power from these multiple signals and will in fact receive different multipaths as the angle of orientation varies.

2.4.1 Polarisation Diversity

In order to improve or extend system performance, some system designers use receive polarisation diversity techniques in an effort to enhance signal reception. In these systems, a circularly polarised or dual linearly polarised antenna is used at the receive site to take advantage of the fact that many linearly polarised multipath signals, with different orientations, exist at the receive site. These dual polarized antennas can accept the orthogonal signals and combine them in the receiver, and so have a greater probability of receiving more total power than a single linearly polarised antenna.

Typically, in polarisation diversity systems, when using a dual linear polarised antenna the receiver samples and tracks the polarisation output providing the strongest signal level (selection combining). Each output will provide a total signal that is a combination of all incident signals arriving in that polarisation. This combined signal will be a function of the amplitude and phase of each signal, as well as the polarisation mismatch of each signal as described by Equation (2.5).

In polarisation diversity systems using a circularly polarised antenna, the receiver only samples the single output. The total signal developed at the output will be a combination of all signals arriving at the antenna. Again this is a function of the individual amplitude and phase of each signal, as well as the polarisation mismatch loss between the circularly polarised and linearly polarised signals as described in Figure 2.19.

The choice between using a circularly polarised antenna or a dual linearly polarised antenna is difficult to determine. This choice is really a function of the make-up of the total signals arriving at the receive site. If the total signal arriving at the receive site is predominantly contained in a linearly polarised wave, then a dual linearly polarised antenna may be the preferred choice. However, antenna alignment is critical in determining the total signal received.

With a dual polarised antenna, the signal loss due to a polarisation mismatch will be between 0 and 3 dB. If a circularly polarised antenna is used (assuming a 1 dB maximum axial ratio over the main beam), the signal loss due to a polarisation mismatch will be between 2.5 and 3.5 dB. If the total signal arriving at the antenna is comprised of a random sample of multiple linearly polarised signals, the circularly polarised antenna will be able to detect the waves, and may be the correct choice for a receiver in a dense scattering environment. With a dual linearly polarised antenna the polarisation mismatch loss will generally be greater than 3 dB.

A number of researchers have shown that dual polarisation diversity, using vertical and horizontal polarisations, can improve the received signal-to-noise ratio [40, 41]. In addition, the performance of a three-branch orthogonal polarisation diversity system in a scattering environment has been investigated and compared to that of a dual-channel polarisation diversity system. The results show that the former system has a 2 dB advantage over the latter [42]. The results also suggest that the use of horizontal polarisation at the transmitter results in a 2 dB improvement over a vertically polarised transmitter. This observation could be explained by the fact that in the indoor environment (where the receive antennas were located), the majority of reflectors are horizontal (floors and ceilings). The work presented showed that there are clear benefits to using a three-branch polarisation diversity scheme to improve the link budget.

Diversity schemes have been shown to be highly efficient in mitigating the effects of multipath fading. The three-branch polarisation diversity schemes [39, 40] can also be applied to millimetre wave antenna systems. This particular scheme uses three orthogonal antennas at the receiver to increase the link budget by more than 6 dB and 2 dB for the Rayleigh and dual-channel cases, respectively. The approach has an added advantage of being relatively small and compact, since the antenna elements can be co-located, making it suitable for applications where space is limited. This ability of the system to provide three uncorrelated copies of the transmitted signal also implies that it can potentially be deployed at both the transmitter and receiver in a conventional multiple-input multiple-output (MIMO) arrangement, to enhance both the capacity and the link budget of the channel. However, in this section emphase will be placed on the operation of the scheme as a diversity system (for a robust channel), while the MIMO analysis will be reported in a future publication. The performance of the triple polar scheme was analysed for the indoor environment in Reference [42]. It was envisaged that users requiring high-speed data services will typically be located in this environment and will be relatively stationary, with data terminals larger than current mobile phones which are capable of accommodating the diversity antennas [43]. The polarisation diversity scheme can then be implemented at the access points and/or at the mobile devices to enhance the link budget.

The three-branch polarisation diversity system employs three orthogonal antennas which may be implemented as either electric or magnetic elements (a total energy antenna can be constructed by using both electric and magnetic elements). In this configuration, three orthogonal electric field detectors were used. One of the antennas is in a vertical position and two are in orthogonal horizontal positions (H1 and H2), as shown in Figure 2.21. As mentioned above, one of the major benefits of this configuration is that the antennas are co-located and can be designed to occupy minimal space. This is particularly important in access points where space is limited, and also in handsets where the device size cannot accommodate spatially separated antennas. One of the problems of having closely separated antennas is the mutual coupling between the elements, which can adversely affect the application of the array in a diversity system (the communications channels then become correlated). However, by careful design, sufficient isolation can be achieved. In [42] and [44], a measurement was conducted in an anechoic chamber to quantify the isolation between the elements. It was observed that, on average, the isolation between the elements is more than 40 dB, which shows that the mutual coupling is negligible and the arrangement is therefore suitable for diversity [42, 44].

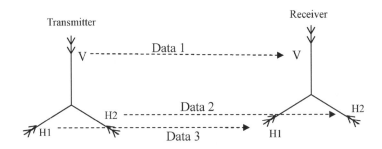

Figure 2.21   Configuration of the three-branch polarisation diversity scheme at transmitter and receiver (V represents vertical polarisation, H1 represents horizontal polarisation 1 and H2 represents horizontal polarisation 2)

References

[1] '60 GHz System', http://dept106.eng.ox.ac.uk/wb/pages/research/microwave/uture-radio-systems/60ghz-imaging.php

[2] A. M. Street, J. G. O. Moss, A. P. Jenkins, D. J. Edwards and M. J. Mehler, 'Indoor Propagation Measurements at Millimetric Frequencies',, ICAP '95, Ninth International Conference on *'Antennas and Propagation'*, Conference Publication 407 , Vol.2 , 4–7 April 1995, pp. 9–12.

[3] Robert S. Elliott, *'Antenna Theory and Design'*, Series on Electromagnetic Wave Theory, IEEE Press, 2006.

[4] G. Von Trentini, 'Partially Reflecting Sheet Arrays' *IEEE Transactions on Antennas and Propagation*, **4**, 1956, 666–671.

[5] D. R. Jackson and N. G. Alexópoulos, 'Gain Enhancement Methods for Printed Circuit Antennas', *IEEE Transactions on Antennas and Propagation*, **33**(9), September 1985.

[6] D. R. Jackson and A. A. Oliner, 'A Leaky-Wave Analysis of the High-Gain Printed Antenna Configuration', *IEEE Transactions on Antennas and Propagation*, **36**, July 1988.

[7] A. P. Feresidis and J. C. Vardaxoglou, 'High Gain Planar Antenna Using Optimised Partially Reflective Surfaces', *IEE Proceedings on Microwave Antennas and Propagation*, **148**, 2001, 345–350.

[8] Alexandros P. Feresidis, George Goussetis, *et al.*, 'Artificial Magnetic Conductor Surfaces and Their Application to Low-Profile High-Gain Planar Antennas', *IEEE Transactions on Antennas and Propagation*, **53**(1), January 2005, 209–215.

[9] M. Thévenot, C. Cheype, A. Reineix and B. Jecko, 'Directive Photonic Bandgap Antennas', *IEEE Transactions on Microwave Theory and Techniques*, **47**, November 1999, 2115–2122.

[10] B. Temelkuaran, M. Bayindir, E. Ozbay, R. Biswas, M. M. Sigalas, G. Tuttle and K. M. Ho, 'Photonic Crystal-Based Resonant Antenna with a Very High Directivity', *Journal of Applied Physics*, **87**, 2000, 603–605.

[11] R. Biswas, E. Ozbay, B. Temelkuran, Mehmet Bayindir, M. M. Sigalas and K.-M. Ho, 'Exceptionally Directional Sources with Photonic-Bandgap Crystals', *Journal of Optical Society of America B*, **18**(11), November 2001, 1684–1689.

[12] A. Fehrembach, S. Enoch, and A. Sentenac, 'Highly Directive Light Sources Using Two-Dimensional Photonic Crystal Slabs', *Applied Physics Letters*, **79**, December 2001, 4280–4282.

[13] C. Cheype, C. Serier, M. Thevenot, T. Monediere, A. Reinex and B. Jecko, 'An Electromagnetic Bandgap Resonator Antenna', *IEEE Transactions on Antennas and Propagation*, **50**(9), September 2002, 1285–1290.

[14] Y. J. Lee, J. Yeo, R. Mittra and W. S. Park, 'Application of Electromagnetic Bandgap (EBG) Superstrates with Controllable Defects for a Class of Patch Antennas as Spatial Angular Filters', *IEEE Transactions on Antennas and Propagation*, **53**(1), January 2005, 224–235.

[15] R. Gardelli, M. Albani and F. Capolino, 'Array Thinning by Using Antennas in a Fabry–Perot Cavity for Gain Enhancement', *IEEE Transactions on Antennas and Propagation*, **54**(7), July 2006.

[16] A. R. Weily, L. Horvath, K. P. Esselle, B. C. Sanders and T. S. Bird, 'A Planar Resonator Antenna Based on a Woodpile EBG Material', *IEEE Transactions on Antennas and Propagation*, **53**(1), January 2005, 216–223.

[17] K. C. Gupta, 'Narrow Beam Antenna Using an Artificial Dielectric Medium with Permittivity Less Than Unity', *Electronic Letters*, **7**(1), January 1971, 16.

[18] I. J. Bahl and K. C. Gupta, 'A Leaky-Wave Antenna Using an Artificial Dielectric Medium, *IEEE Transactions on Antennas and Propagation*, **AP-22**, January 1974, 119–122.

[19] S. Enoch, G. Tayeb, P. Sabouroux, N. Guérin and P. Vincent, 'A Metamaterial for Directive Emission,' *Physical Review Letters*, **89**, November 2002, 213902-1–213902-4.

[20] G. Lovat, P. Burghignoli, F. Capolino, D. R. Jackson and D. R. Wilton, 'Analysis of Directive Radiation from a Line Source in a Metamaterial Slab with Low Permittivity', *IEEE Transactions on Antennas and Propagation*, **54**(3), March 2006, 1017–1030.

[21] G. Lovat, P. Burghignoli, F. Capolino, *et al.*, 'High Directivity in Low-Permittivity Metamaterial Slabs: Ray-Optic vs Leaky-Wave Models', *Microwave and Optical Technology Letters*, **48**(12), December 2006, 2542–2548.

[22] G. Lovat, P. Burghignoli, F. Capolino and D. R. Jackson, 'On the Combinations of Low/High Permittivity and/or Permeability Substrates or Highly Directive Planar Metamaterial Antennas', *IEE Proceedings on Microwave Antennas and Propagation*, Special Issue on Metamaterials, Vol. 1, February 2007.

[23] G. Lovat, P. Burghignoli, F. Capolino and D. R. Jackson, 'Highly-Directive Planar Leaky-Wave Antennas: A Comparison between Metamaterial-Based and Conventional Designs', *EuMA (European Microwave Association) Proceedings*, Vol. 2, 2006.

[24] K. Huang and Z. Wang 'V-Band Patch-Fed Rod Antennas for High Data-Rate Wireless Communications', *IEEE Transactions on Antennas and Propagation*, **54**(1), January 2006, 297–300.

[25] W. Menzel, D. Pilz and M. Al-Tikriti, 'Millimeter-Wave Folded Reflector Antennas with High Gain, Low Loss, and Low Profile', *IEEE Antennas and Propagation Magazine*, **44**(3), June 2002, 24–29.

[26] F. Colomb, K. Hur, W. Stacey and M. Grigas, 'Annular Slot Antennas on Extended Hemispherical Dielectric Lenses', Antennas and Propagation Society International Symposium, 1996, *AP-S Digest*, 3, July 1996, 2192–2195.

[27] S. Sierra-Garcia and J.-J. Laurin, 'Study of a CPW Inductively Coupled Slot Antenna', *IEEE Transactions on Antennas and Propagation*, 47(1), January 1999, 58–64.

[28] W. Y. Ali-Ahmad, G. V. Eleftheriades, L. P. B. Katehi and G. M. Rebeiz, 'Millimeter-Wave Integrated-Horn Antenna. II. Experiment', *IEEE Transactions on Antennas and Propagation*, **39**(11), November 1991, 1582–1586.

[29] J. A. Besley, N. N. Akhmediev and P. D. Miller, *Optics Letters*, **22**(15), 1 August 1997.

[30] X. Wu, G. V. Eleftheriades and T. E. van Deventer-Perkins, 'Design and Characterization of Single- and Multiple-Beam mm-Wave Circularly Polarized Substrate Lens Antennas for Wireless Communications', *IEEE Transactions on Microwave Theory and Techniques*, **49**(3), March 2001, 431–441.

[31] M. Sironen, Y. Qian and T. Itoh, 'A 60 GHz Conical Horn Antenna Excited with Quasi-Yagi Antenna', *Microwave Symposium Digest, IEEE MTT-S International*, **1**, 2001, 547–550.

[32] Xidong Wu and G. V. Eleftheriades, 'Two-Lens and Lens-Fed Reflector Antenna Systems for mm-Wave Wireless Communications', IEEE Antennas and Propagation Society International Symposium, 2000, Vol. 2, pp. 660–663.

[33] L. J. Chu, 'Physical Limitations of Omni-directional Antennas', *Journal of Applied Physics*, **19**, December 1948, 1163–1175.

[34] R. F. Harrington, 'On the Gain and Beamwidth of Directional Antennas', *IRE Transactions on Antennas and Propagation*, **6**, 1958, 219–225.

[35] W. Geyi, 'Physical Limitations of Antenna', *IEEE Transactions on Antennas and Propagation*, **51**(8), August 2003, 2116–2123.

[36] R. F. Harrington, 'Effect of Antenna Size on Gain, Bandwidth, and Efficiency', *Journal of Research National Bureau of Standards – D. Radio Propagation*, **64D**(1), January/February 1960.

[37] A. C. Ludwig, 'The Definition of Cross Polarization, *IEEE Transactions on Antennas and Propagation*, **AP-21**(1), January 1973, 116–119.

[38] R. L. Fante, 'Quality Factor of General Ideal Antennas', *IEEE Transactions on Antennas and Propagation*, **AP-17**(2), March 1969, 151–155.

[39] T. Manabe *et al.*, 'Polarization Dependence of Multipath Propagation and High-Speed Transmission Characteristics of Indoor Millimeter-Wave Channel at 60GHz', *IEEE Transactions on Vehicular Technology*, **44**(2), May 1995.

[40] L. Lukama, K. Konstantinou and D.J. Edwards, 'Polarization Diversity Performance for UMTS', Proceedings of the International Conference on *'Antennas and Propagation'* (ICAPZOOI), April 2001, Manchester, England.

[41] R.G. Vaughan, 'Polarization Diversity in Mobile Communications', *IEEE Transactions on Vehicular Technology*, **39**(3), August 1990, 177–186.

[42] L. C. Lukama, D. J. Edwards and A. Wain, 'Application of Three-Branch Polarisation Diversity in the Indoor Environment', *IEE Proceedings on Communications*, **150**(5), October 2003, 399–403.

[43] L.C. Lukama, K. Konstantinou and D.J. Edwards, 'Performance of a Three-Branch Orthogonal Polarization Diversity Scheme', *IEEE Vehicular Technology Conference Proceedings*, Fall, 2001, 2033–2037.

[44] L. Lukama, K. Konstantinou and D.J. Edwards, 'Three-Branch Orthogonal Polarization Diversity Scheme', *Electronic Letters*, **37**(20), 2001, 1258–1259.

3

Planar Antennas

There has been rapid growth in printed antenna theory and technology during the last decade. Characteristics of printed antennas, such as low-cost, low-profile, conformability and ease of manufacture, have been studied and the advantages have been shown to outweigh the electrical disadvantages: such as narrow bandwidth and low-power capacity, for certain applications.

This chapter will discuss properties of planar antenna elements suitable for millimetre wave systems, and presents data on their electrical characteristics and other features relevant to this application. Section 3.1 focuses on printed dipoles and patches, describes their basic features and gives some simple design rules. Section 3.2 presents the basic concepts of slot antennas, while Section 3.3 presents a basic description of a quasi-Yagi antenna. Section 3.4 discusses wideband bowtie antennas, and Section 3.5 introduces reflector antennas. Section 3.6 describes design and test considerations. Finally, Section 3.7 discusses production and manufacturing issues

3.1 Printed Antennas

Printed dipole radiating elements have been extensively studied by many researchers. Essentially the configuration is a planar dipole element supported on a dielectric substrate. A useful discussion of these antennas was reported by Alexopoulos and Rana in Reference [1], where an analysis was presented using basically a method of moments procedure. Figure 3.1 shows a typical centre-fed printed dipole element. The use of parallel stripline feed lines to couple the radiating dipole, as in Reference [2], can alleviate the feeding difficulty at the expense of a more complicated feeding structure, possibly involving printed conductors on two substrate levels. Advantages of the printed dipole are that it uses less substrate area compared to patch elements (see below) which is particularly important in arrays, and that it can be used near its first or second resonant frequencies without deleterious higher-order modes being excited.

The printed patch antenna in Figure 3.1 can be fed with a microstripline or with a probe connecting the radiating element through an entry hole from the bottom of the substrate. In practice, a coaxial feed would have its outer conductor joined to the ground plane and the inner

Millimetre Wave Antennas for Gigabit Wireless Communications   Kao-Cheng Huang and David J. Edwards
© 2008 John Wiley & Sons, Ltd

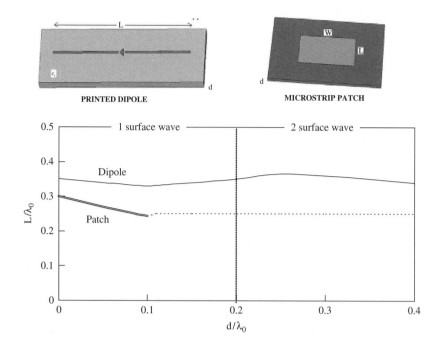

Figure 3.1   Resonant lengths of a printed dipole and a microstrip patch versus d for polystyrene substrate. Patch width $W = 0.3\,\lambda_0$. At the top are shown the printed dipole and patch geometries. (Reproduced by permission of © 1983 IEEE [3])

conductor, or the feeder, would pass through the dielectric substrate and connect to the radiating element. It is observed that the rectangular patch needs more substrate area than the dipole and that a probe-type feed may be difficult to fabricate on monolithic substrates, such as alumina or quartz substrates. A feeding issue exists with microstripline feeds, since for a microstripline, the line width determines its characteristic impedance (for a given substrate) and is relatively constant with frequency. The size of a resonant patch antenna, however, decreases with increasing frequency, so that a given microstrip feed line on a substrate has an effective upper frequency limit beyond which the resonant patch width would be narrower than the feed line width.

It should be noted that there are intrinsic differences in the electrical operation of printed dipoles and patches. The first resonance of a printed dipole, like a half-wave dipole in free space, is a *series-type resonance* as the current is driven in the same direction as the driving electric field. On the other hand, the first resonance of a patch antenna is a *parallel-type (anti)resonance* as the current is in the plane of the patch but the field is between the conducting plates of the structure. This difference is a result of the field structure created in the vicinity of the element by a particular feed.

For a dipole, the feed couples to the electric field component along the dipole axis, while a coaxial or microstripline-fed patch antenna is coupled by the electric field component perpendicular to the substrate. Printed dipoles and patches, however, have similar current distributions; thus the radiation patterns are similar [1, 4].

For a nearly square patch with dimensions a and b, the design procedure can be listed as follows:

(i) Find the resonant frequency from the effective patch dimensions a and b, satisfying:

$$a \geq \frac{c}{2f\sqrt{\varepsilon_r}} \geq b$$

where c is the speed of light and ε_r is the relative dielectric constant of the material. The true physical dimensions for W and L are smaller than a and b because of the fringing effect.

(ii) Choose the operation frequency and bandwidth.

(iii) Choose the appropriate substrate.

(iv) Estimate the power loss.

(v) Estimate the resonant resistance and quality factor.

(vi) Determine the feeding method.

(vii) Occasionally, iterations are needed for fine-tuning the antenna performance.

These points are discussed step by step as follows.

(i) Resonant Frequency
The first consideration in the design of a printed antenna element is the length L of the element required for resonance. This length is a function of substrate thickness d and the relative dielectric constant ε_r, and, in the case of a microstrip patch, a function of the patch width W. Because the dielectric material only exists in the region of space underneath the antenna element and not in the space above, the resonant length does not scale with the dielectric constant as $1/\sqrt{\varepsilon_r}$ (which is only used by an antenna in a homogeneous medium).

Figure 3.1 shows the required lengths for the first resonance of a printed dipole and a rectangular microstrip patch element versus substrate thickness d, for a Teflon ^{TM} material. The patch width is $W = 0.3\lambda_0$ (λ_0 is the free space wavelength). The dipole length varies less than 6 % for $0 < d < 0.5b$, and is slightly longer than the patch length, which varies proportionately more with d. An interesting feature of the patch antenna is that it stops resonating for substrate thicknesses greater than $0.11\lambda_0$. With increasing substrate thickness, the trend is towards an entirely inductive input impedance locus (due to the phase relationship between current and applied field/voltage). This effect occurs for both probe-type and microstripline-type feeds.

This situation is probably undesirable in most cases, and so the use of patches on thick substrates may not be practical unless some way of countering this inductive trend is used, for example by using a capacitive-gap coupling from a feed line. The dotted continuation of the patch length curve in Figure 3.1 is given only to show the length chosen for the calculation of other data presented later in Figure 3.2. An increase in patch width W can reduce the resonant length by a few per cent; the length reduction is greater for thicker substrates.

(ii) Frequency Bandwidth

Frequency bandwidth is one of the important antenna parameters. It is defined as the half-power width of the equivalent circuit impedance response. For a series-type resonance [2], this bandwidth (BW) is:

$$BW = \frac{2R}{\omega_0 \, \mathrm{d}X/\mathrm{d}\omega|_{\omega_0}} \tag{3.1}$$

where the input impedance at the resonant frequency ω_0, Z, is equal to $R + \mathrm{j}X$. For a parallel-type resonance, Equation (3.1) is used with R replaced by the conductance G, and X replaced by the inverse of the reactance B; here the input admittance Y at resonance is equal to $G + \mathrm{j}B$. The derivative in Equation (3.1) can be evaluated by calculating the input impedance at two frequencies near resonance and using a finite difference approximation.

As discussed by Harrington [5], the lowest achievable Q of an antenna is inversely related to the antenna volume (since Q depends on the ratio of the stored energy [depends on L^3, the volume] and to the dissipated energy [linear in L or L^2, the length or area]). Since the patch antenna encompasses a greater volume than the printed dipole, its Q can be lower than the Q of the dipole, and hence the operational bandwidth can be greater.

(iii) Substrate Characteristics

To produce printed antennas, it is necessary to select a material that is mechanically and electromagnetically stable. This choice of substrate is based on the fact that the permittivity range is from 1.2 to 13 and that these materials are either in use today or are expected to be in use for millimetre wave antenna systems in the future. Because of space considerations, not all materials available today will be compared here, but it is felt that these examples are representative and other common substrate materials have properties roughly in the range of those considered here.

Teflon and related products like Rexolite™ and RT/Duroid 5870/5880™ have been used extensively in the microwave band [6]. Quartz substrates have very good dimensional stability and are often used in microwave integrated circuits. Gallium arsenide is one of the low-loss substrate materials used for monolithic microwave integrated circuits. DuPont™ 943 Low Loss Green Tape™ [7] can be used with gold, silver and mixed-metal material systems, and can be considered for use by millimetre wave circuit designers and manufacturers.

The dielectric loss factor, also known as the dissipation factor, is defined as the tangent of the loss angle (tan δ, or the loss tangent). The loss factor represents the ratio of resistance to reactance of a parallel equivalent circuit of the ceramic element. Alternatively, for a primarily reactive circuit model, the degree to which the angle differs from 90° would be a case of a lossless device. Thus the real and imaginary components of the dielectric constant are quantified. The dielectric constant (ε) and loss factor (tan δ) can be measured using a standard impedance bridge or an impedance analyser. Some typical examples are identified in Table 3.1.

(iv) Power Loss

Power loss can be caused by surface waves, the dielectric material, and implementation tolerances, etc. Here the first two losses will be discussed:

Table 3.1 Electrical properties of microwave substrates [8]

	loss tangent

permittivity	0.0001	0.0002	0.0003	0.0004	0.0005	0.0006	0.0007	0.0008	0.0009	0.001	0.002	0.005	0.010	0.050	0.100
1.0															
1.2														MgCO3	
2															
2.1	Teflon														
2.2			Polypropylene												
2.3	Polyethylene														
2.4															
2.5		Polystyrene													
3	Quarz														
4			BN												
5			Mica												
6				BeO											
7											GreenTape943				
8															
9			Sapphire	MgO		Al2O3									
10													MgTiO3		
12.5										GaAs					

- Power loss due to surface waves

Both TE and TM surface waves can be excited on a grounded dielectric substrate. The cut-off frequency of these modes is given by [5]:

$$f_c = \frac{nc}{4d\sqrt{\varepsilon_r - 1}} \qquad (3.2)$$

where c is the speed of light and $n = 0, 1, 2, 3$ for the TM_0, TE_1, TM_2, TE_3 surface modes, respectively. Note that the TM_0 mode has a zero cut-off frequency, so that it can be generated for any substrate thickness. As the substrate becomes electrically thicker, more surface modes can exist and the coupling to the lower-order modes can become stronger. For thin substrates ($d < 0.01 \lambda_0$) this surface wave excitation is generally regarded as unimportant.

For thicker substrates these surface waves may have a detrimental effect on printed antenna performance. Surface wave power launched in an infinitely wide substrate, would not contribute to the main beam radiation and so can be considered as a loss mechanism. In this case, the radiation efficiency can then be defined as:

$$e = \frac{P_{\text{rad}}}{P_{\text{rad}} + P_{\text{sw}}} \qquad (3.3)$$

where P_{rad} is the power radiated via the space wave (the main beam power), and P_{sw} is the power coupled into surface waves. $P_{rad} + P_{sw}$ is then the total power delivered to the printed antenna element. Here dielectric losses have been ignored. The effect of a finite-sized substrate is to diffract the surface wave energy from the substrate edges, possibly causing undesirable effects on sidelobe level, polarisation or main beam shape. In general, for low sidelobe performance, the edge excitation needs to be low and the aperture excitation needs to vary slowly over the patch. Low sidelobe requirements tend to conflict with high aperture efficiency. However, by a careful balancing of the modal excitation, a good compromise between mainlobe gain and sidelobe suppression can be achieved. A high current excitation over as much of the patch as possible (efficiency) is aimed for, but with the current tapering smoothly to zero at the edges. Surface waves could also be diffracted by or coupled to feed lines or components on the substrate.

In the moment method formulation, surface wave fields and space wave fields are easily separated from the Sommerfeld-type integral expression for the total fields of an elemental current source on a grounded dielectric slab. The surface waves arise from the residues of the contour integral. Since in the moment method the impedance matrix elements are expressed in terms of integrals of the fields from the modal expansion, these elements can be separated, as:

$$Z_{mn} = Z_{mn}^{rad} + Z_{mn}^{sw} \tag{3.4}$$

where Z_{mn} represents the matrix element using the total field, and Z_{mn}^{rad} and Z_{mn}^{sw} represent the direct radiation (the space wave) and the surface wave contributions, respectively. Then, if I_n represents the current on the nth expansion mode, the total input power can be written as:

$$P_{tot} = \text{Re} \sum_n \sum_m I_n^* Z_{mn} I_m \tag{3.5}$$

and the radiated power can be written as:

$$P_{rad} = \text{Re} \sum_n \sum_m I_n^* Z_{mn}^{rad} I_m \tag{3.6}$$

- Losses due to the dielectric

Power loss due to dielectric heating can be calculated by using the loss tangent and complex permittivity for the particular substrate dielectric material. For the half-wave dipole (a series-type resonance), for example, the radiation efficiency based on the dielectric loss can be calculated as:

$$\eta = \frac{R_r}{R_r + R_l} \tag{3.7}$$

where R_r is the radiation resistance at the input terminals and R_l is the loss resistance. R_r and R_l can be found from the two calculations of the input impedance, one with $\tan \delta = 0$ and one with $\tan \delta \neq 0$. The radiation resistance is $R = Re(Z_{in})$, for $\tan \delta = 0$ and the loss resistance is found from $R_r + R_l = Re(Z_{in})$ with $\tan \delta \neq 0$. This is an accurate procedure for small losses. For full-wave dipoles or microstrip patches (anti-resonances), the efficiency is calculated using the conductance in Equation (3.7). Note that efficiency as defined by Equation (3.7) does not include the power loss due to surface waves (although it does in fact include heating loss from surface wave fields).

It has been proven that the patch efficiency is greater than the dipole efficiency, and that the efficiency improves rapidly as the substrate thickness increases [3]. Both of these effects can be explained by noting that, for a given power level, the fields are more concentrated for thin substrates or narrow antenna elements; thus more power is lost to dielectric heating than in cases of thicker substrates or wider elements.

(v) Resonant Resistance

Having introduced the quality factor in Section 2.3, this subsection will concentrate on resonant resistance.

Figure 3.2 shows the input resistance of a half-wave printed dipole and a microstrip patch on a Teflon substrate versus its thickness. As previously pointed out, the patch element does not strictly resonate for the substrate thickness $d > 0.11 \lambda_0$; the patch resistance shown in Figure 3.2 for the substrate thickness $d > 0.11 \lambda_0$ is the real part of the input impedance for a patch length of $0.270 \lambda_0$. Since the printed dipole's first resonance is a series-type resonance, the input resistance is very small for small d, since electrically thin substrates imply high Q resonance. The microstrip patch, having a parallel-type resonance, shows a high input resistance for small d.

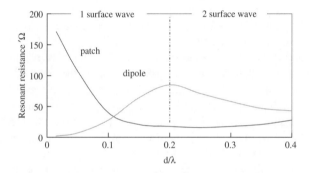

Figure 3.2 Input resistance of a half-wave printed dipole and a microstrip patch versus d for Teflon. The patch is probe fed at a point $L/4$ from the edge and the patch width $W = 0.3 \lambda_0$. (Reproduced by permission of © 1983 IEEE [3])

The full-wave dipole has a parallel-type resonance, with high input resistance for small substrate thickness d, similar to a full-wave dipole in free space. This element has interesting advantages in some applications [9]. First, its half-power beamwidth is significantly less than that of a half-wave dipole. Second, if a pair of full-wave dipoles are arranged $\lambda_0/2$ apart to form a subarray element, the E and H plane beamwidths will be about equal, and if a detector diode is placed in the centre of the subarray and connected to the dipoles by a printed parallel line, as in Reference [9], the $\lambda_0/4$ line length will yield an impedance inversion from the high input resistance of the full-wave dipoles to a low impedance matching to the diode.

All the dipoles are centre fed and all the patches are probe fed at a point $L/4$ from the (radiating) patch edge. Moving the feed position towards the end of the dipole or patch will increase the input resistance, at the first resonance.

(vi) Feeding Methods

Six basic methods for feeding patch antenna are shown in Figure 3.3. The slotline feed, coplanar waveguide feed, and aperture-coupled feed configurations have attracted much attention because of their suitable geometries for monolithic integration. Figure 3.3 (a) shows the probe feed via hole method. Its advantage is that there is no feed line radiation loss, and little coupling between the patch and feed line (the currents are orthogonal). Also, the impedance of this patch can be accurately and easily predicted and a different value of impedance can be obtained by choosing the feed location. However, the fabrication can be complicated and costly for millimetre wave applications. Figure 3.3 (b) shows a microstripline edge feed method, which can simplify the fabrication process as the antenna and the feeding lines are printed in one step. Its drawback is its inflexibility in design, since both feed and patch are over the same substrate, resulting in possible erratic radiation coupling for millimetre waves. Figure 3.3 (c) shows the microstripline sandwich feed method, which is flexible in microstripline and patch design. However, two layers of substrates are required and it can be difficult to integrate them with active devices due to their heat dissipation. Figure 3.3 (d) shows a slotline feed method, which is simple to fabricate and easy to integrate with active devices. Also it is simple to allow for heat dissipation, and it is possible to etch the patch and the slot in one step. However, some possible stray radiation may be generated by the slot and there is limited flexibility in a large feeding network layout. Figure 3.3 (e) shows a coplanar waveguide feed method, which again is simple to fabricate and easy to integrate with active devices. The transitions to active devices and MMICs are simple, and only a small amount of stray radiation comes from feed. This method requires more space and has less freedom when designing large feed networks. Figure 3.3 (f) shows an aperture-coupled feed method. In this configuration, there is an aperture in the ground plane, which allows electromagnetic coupling between the patch and the feed line. This method has more design freedom, as the feeding network and patches can be designed separately to a large extent. However, this method is costly and complex compared to other methods as it requires multiple layers of conductor and substrate. Also more space under the ground plane is required. Generally, in the aperture-coupled patch antenna (Figure 3.4), the thickness of the ground plane corresponds to the thickness of the metallisation substrates.

At 60 GHz, in order to make the structure rigid and to introduce active components that may be associated with the feeding line network, it is interesting to consider increasing the thickness of the ground plane. Two different technologies, one with the Duroïd substrate and the other one with TPX (TPX® is a registered trademark of Mitsui Plastics), have been developed to introduce a thick copper ground plane between the two distinct substrate layers. With the Duroid substrate (Figure 3.5), the initial copper film is first removed and then the two substrate layers are bonded and pressed at high temperature. For the TPX realisation, several sheets of substrate are stacked and pressed at high temperature to obtain the desired TPX thickness.

To illustrate these technologies, two aperture-coupled microstrip patch antennas, calculated by an extension of the cavity method [11], have been produced. With these two antennas, the same radiation efficiency of around 70 % is obtained. These prototypes show that these technologies are very suitable for millimetre waves. An example of a circularly polarised patch antenna is represented in Figure 3.6 (a). The circular polarisation is due to the cross apertures in the ground plane and the tilted feeding line, which allows the excitation on two orthogonal modes in the patch. This antenna is realised with a glass Teflon technology. The

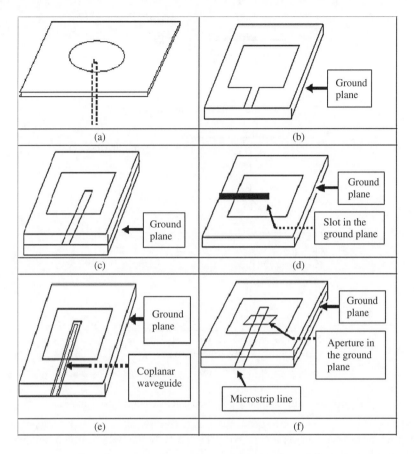

Figure 3.3 (a) Probe feed, (b) microstripline edge feed, (c) microstripline sandwich feed, (d) slotline feed, (e) coplanar waveguide feed and (f) aperture-coupled feed [10]

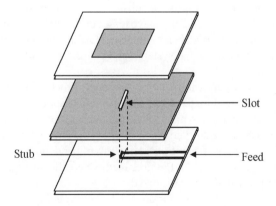

Figure 3.4 Aperture-coupled patch antenna

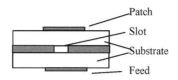

Figure 3.5   Glass Teflon technology

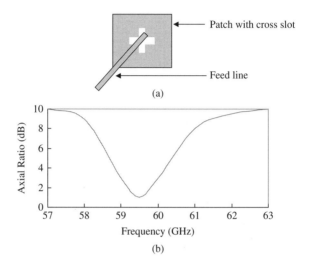

Figure 3.6   (a) Circularly polarised antenna and (b) its axial ratio

measured axial ratio is given in Figure 3.6 (b). The best circular polarisation is obtained around 59.7 GHz and the axial ratio is equal to 1 dB.

Printed antennas are the obvious choice for integration in a stacked configuration, due to their low profile and planar characteristics. However, microwave and optical devices are commonly fabricated on high dielectric constant substrates, such as gallium arsenide. The design of printed antenna elements on such substrates is commonly avoided, as they suffer from a narrow bandwidth ($< 6\,\%$) and excessive losses due to surface waves, which decreases the overall gain of the antenna. Wideband antenna elements are desirable to ensure that the antenna is not the limiting factor in the system bandwidth, and to allow for the possibility of multiservice transmission.

To increase the frequency bandwidth and/or efficiency, this section presents a comparison of two different broadband millimetre wave antenna structures fabricated on high dielectric constant substrates. The geometries under investigation are the hilo stacked patch and the coplanar waveguide fed aperture stacked patch (ASP) [12]. The relative merits and shortcomings of these structures will now be highlighted in terms of bandwidth, surface wave loss, back radiation and ease of integrated design. Possible methods to alleviate the shortcomings of the individual configurations will also be postulated.

The general configuration of the hilo stacked patch is given in Figure 3.7 (c). For the current purpose, the feed substrate has the relative permittivity of gallium arsenide ($\varepsilon \sim 12.9$) and the antennas are designed to operate in the millimetre wave band. The feed wafer thickness is

240 μm for the hilo stacked patch. Gold metallisation is used on the feed substrate. Square patch elements are employed in both antenna structures to accommodate the further development of circularly/dual polarised versions.

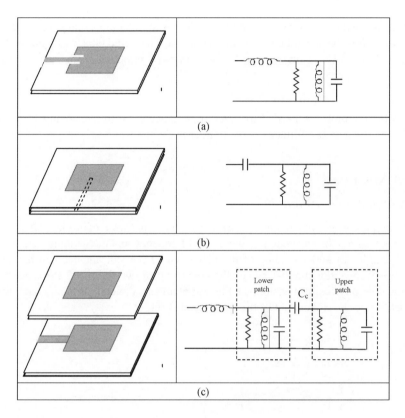

Figure 3.7   Structure of the millimetre wave patch antenna and the equivalent circuit: (a) inset microstrip feed, (b) proximity coupled feed and (c) stacked patch [13]

The circuit model for the stacked patch antenna can be developed from physical considerations using the cavity method, and an expression of input impedance is obtained as a function of antenna parameters and frequency. The equivalent circuit, as shown in Figure 3.7 (c), consists of two cascaded parallel resonant circuits with a series coupling capacitance. The lower patch can be modelled as a parallel *RLC* circuit with a capacitor C_c in series, representing the cross coupling between the upper patch and the lower patch. An inductance L_p is connected as a direct feed.

If considering the feeding patch to be excited in the TM_{11} mode and the radiating patch is operated in the TM_{11} mode, then the value of *RLC* for the TM_{11} mode can be found as:

$$R = \frac{Qh}{\pi f \varepsilon_r \varepsilon_0 l_{\text{eff}} w_{\text{eff}}} \cos^2\left(\frac{\pi x_i}{l}\right) \cos^2\left(\frac{\pi y_i}{w}\right) G_{mn}$$

$$C = \frac{Q}{2\pi f_r R}$$

$$L = \frac{R}{2\pi f_r Q}$$

where Q is the quality factor associated with all losses in radiation, conductor, dielectric and surface waves, h is the substrate height, l_{eff} and w_{eff} are the effective length and width of the patch, respectively, $G_{mn} = \sin C (m\pi d_p/2l) \sin C(n\pi d_p/2w)$ and d_p is the effective width of a uniform feeding patch.

(vii) Iterations
Occasionally, iterations are needed for fine-tuning antenna performance and for improving manufacture tolerance at millimetre wave frequency.

3.2 Slot Antennas

3.2.1 Standard Slot Antenna

The slot antenna consists of a radiator formed by cutting a narrow slot in a large metal surface. The slot length is a half-wavelength at the desired frequency and the width is a small fraction of a wavelength. An understanding of the behaviour of the slot antenna can be gained by considering a conventional half-wave dipole consisting of two flat metal strips, as shown in Figure 3.8. The physical dimensions of the complementary metal strips are such that they would just fit into the slot cut out of the large metal sheet.

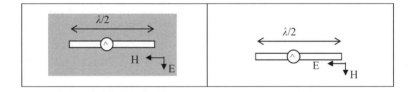

Figure 3.8   Slot antenna (left) and complementary dipole (right)

The slot antenna is compared to its complementary dipole to illustrate that radiation patterns produced by a slot antenna cut into an infinitely large metal sheet and those of the complementary dipole antenna are the same. However, several important differences exist between the slot antenna and its complementary antenna. First, the electric and magnetic fields are interchanged. As a result, the polarisation of the radiation produced by a horizontal slot is vertical. If a vertical slot is used, the polarisation is horizontal. A second difference between the slot antenna and its complementary dipole is that the direction of the lines of electric and magnetic force abruptly reverse from one side of the metal sheet to the other. In the case of the dipole, the electric lines have the same general direction while the magnetic lines form continuous closed loops. When energy is applied to the slot antenna, currents flow in the metal sheet. These currents are not confined to the edges of the slot but rather spread out over the sheet. Radiation, then, takes place from both sides of the sheet. In the case of the complementary dipole, however, the currents are more confined, so a much greater magnitude of current is required to produce a given power output using the dipole antenna.

The general principle of complementary radiators was first identified by Babinet for optics [14]. The concept of complementary radiators is usually referred to as Babinet's principle, which shows that the slot will have the same radiation pattern as a dipole with the same dimensions as the slot, except that the E- and H-fields are interchanged, as illustrated in Figure 3.8, which shows that the slot is a magnetic dipole rather than an electric dipole. As a result, the polarisation is rotated 90°, so that radiation from a vertical slot is polarised horizontally. For instance, a vertical slot has the same pattern as a horizontal dipole of the same dimensions, and so it is possible to calculate the radiation pattern of a dipole. Thus, a longitudinal slot in the broad wall of a waveguide radiates just like a dipole perpendicular to the slot. A fuller discussion of the concept of complimentary radiators can be found in Reference [3].

Slot antenna arrays seem to be one of the good candidates for 60 GHz applications when microstripline feeding systems with a reflecting plate are used. If feeding is on the opposite side of the metallised structure between the substrate and the reflector, the feeding network theoretically will not disturb the radiation pattern. The two types of slot structures will now be examined; one with a double spiral-like element and the second with a V-shape configuration of slot elements. In Figure 3.9 (a) a typical spiral-like slot element is illustrated, and in Figure 3.9 (b) V-slot elements are shown. Both of the approaches are theoretically capable of providing between 15 and 20 % of the bandwidth (at a VSWR of less than 2), and about 8 % for an axial ratio better than 3 dB.

Figure 3.9 (a) Spiral slot antenna prototype for circular polarisation and (b) V-slot antenna prototype for circular polarisation [15]

The S_{11} of the V-slot antenna is shown in Figure 3.10. Its E-plane radiation pattern at 60 GHz for left-hand circular polarisation and right-hand circular polarisation are shown in Figure 3.11. From the result, it is found that the V-slot antenna is circularly polarised and its axial ratio can be tuned by using different slot lengths in the V shape.

3.2.2 Tapered Slot Antennas

Typically these antennas (Figure 3.12) are printed on a thin dielectric substrate and with their axial length being 3 to $12\lambda_0$. Correctly designed, these tapered slots provide good electrical performance, including a gain of 10–17 dB, relatively low sidelobes, a circular symmetric

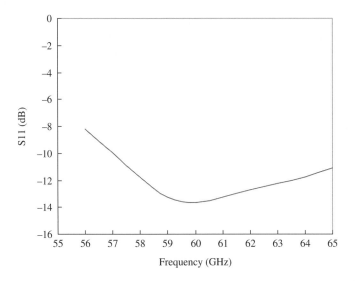

Figure 3.10 S_{11} of the V-shape slot antenna

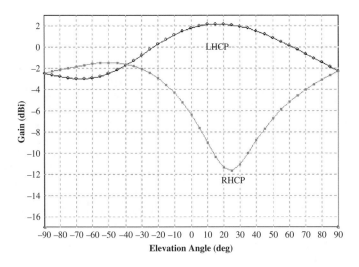

Figure 3.11 E-plane radiation pattern gain display of the V-shape slot antenna at 60 GHz when $\phi = 0°$ for both left-hand circular polarisation (LHCP) and right-hand circular polarisation (RHCP)

main beam and practically constant impedance over a broad frequency band [16]. Achieving circular polarisation or dual polarisation, however, is a problem that is not easily solved. These antennas have been studied, in particular for array applications [16, 17]. Since tapered slot antennas are *endfire* devices, their directivity gain is primarily determined by their axial length when their cross-section width is small. Their main radiation direction is along the axis of the taper. Hence, when used as array elements, they can be packed closely together. Although this will reduce their directivity, their port-to-port isolation tends to remain fairly high. Since, in addition, these antennas are amenable to integration of monolithic solid-state devices, such as Schottky diodes or superconductor–insulator–superconductor (SIS) mixers, and their depth dimension provides ample room for integrated circuits, they are well suited for the design of feed arrays for high- resolution millimetre wave imaging systems and multibeam satellite communication systems [16]. Other applications include scanned arrays for tracking systems and integrated arrays for quasi-optical power combining [17].

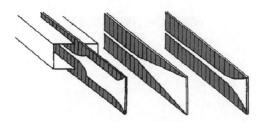

Figure 3.12   Printed tapered slot antennas: (left) antenna with constant width section, (middle) linear taper and (right) Vivaldi antenna (exponential taper). (Reproduced by permission of © 1992 IEEE [18])

3.3 Quasi-Yagi Antennas

This section introduces a quasi-Yagi antenna with a truncated microstrip ground plane as a reflecting element, thus eliminating the need for a reflector dipole. It is a very compact design ($< 0.5\lambda_0 \times 0.5\lambda_0$ for the entire substrate) compatible with any microstrip-based monolithic microwave integrated circuit (MMIC). The quasi-Yagi antenna has several advantages over traditional wire antennas radiating in free space. First, the presence of the substrate provides mechanical support for the antenna and planar transmission line compatibility. Wire-type antennas in free space are naturally fragile at high frequencies and difficult to feed. Second, use of a high-permittivity substrate means that the antenna will be extremely compact in terms of free space wavelengths. A centre-to-centre array spacing of a half-wavelength can be made with this antenna, which corresponds to a free space wavelength at the centre frequency of the antenna. Tighter spacing between elements may be achieved at the cost of increased mutual coupling. In this section, the design and performance of a broadband quasi-Yagi antenna will be presented.

Figure 3.13 shows the layout of the uniplanar quasi-Yagi antenna. The antenna can be fabricated on a single dielectric substrate with metallisation on both sides. The top metallisation consists of a microstrip feed, a broadband microstrip-to-coplanar stripline (CPS) balun, previously reported in Reference [20], and two dipole elements, one of which is the driver element fed by the coplanar stripline, and the second is the parasitic director. The metallisation on the bottom plane is a truncated microstrip ground, which serves as the reflector element for the antenna. The parasitic director element on the top plane simultaneously directs the antenna propagation towards the endfire direction and acts as an impedance-matching parasitic element.

One of the most unique and effective features of this antenna is the use of the truncated ground plane as an ideal reflector, which is completely cut off in the grounded dielectric slab region. The driven printed dipole is used to generate a surface wave with very little undesired content [21], which can contribute to cross polarisation. The dipole elements of the quasi-Yagi antenna are strongly coupled by the surface wave, which has the same polarisation and direction as the dipole radiation fields.

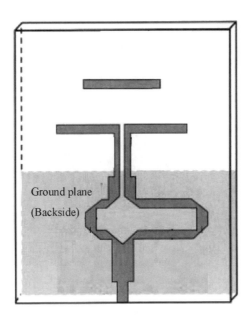

Figure 3.13   Schematic of the quasi-Yagi antenna. (Reproduced by permission of © 2000 IEEE [19])

The antenna has a broad bandwidth (measured 48 % for a VSWR of 2), good radiation profile (front-to-back ratio of 12 dB, cross polarisation of 12 dB) and moderate gain (3–5 dBi) [20]. It should be noted that the pattern is maintained across the entire matched bandwidth.

As with the conventional Yagi–Uda antenna, the design requires careful optimisation of the driver, director and reflector parameters, which include element spacing, length and width. While it may seem counter-intuitive that a broadband antenna will require careful optimisation, this is essential if desirable radiation characteristics are to be maintained across the entire operating bandwidth. Therefore, the bandwidth is defined not only in terms of its matched characteristics but also in terms of radiation characteristics such as cross polarisation, front-to-back ratio and relatively flat gain. It is also found that the choice of substrate is critical for the performance of the antenna. The design requires a high-permittivity design with moderate thickness. This is because the fundamental operation of the antenna relies on surface wave effects, which are strongly dependent on the chosen substrate. For a dielectric with a permittivity of 10.2, a thickness of 0.635 mm can be a good option for millimetre wave operation. When scaling the antenna to other frequency bands, the thickness of the antenna must also be scaled accordingly.

3.4 Bow-Tie Antennas

The bow-tie antenna is another name for a fan dipole antenna, as shown in Figure 3.14. By using triangular elements instead of rectangular, the bandwidth is wider than the patch antenna and can cover the whole 60 GHz ISM band. The bandwidth of these antennas depends on the length of their arms. The centre angle and the length of the bow-tie arms specify the lower frequency [22].

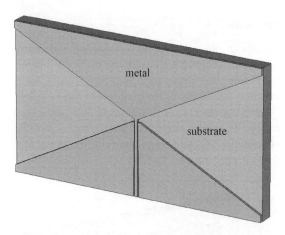

Figure 3.14 Coplanar waveguide feed bow-tie antenna

Bow-tie antennas can be fed by coplanar waveguides, which has advantages such as ease of connection to the surface when mounting active components [23]. Because of the popularity of the coplanar waveguide line for integration with active devices [24, 25], this transmission line has been selected for the feed line of the bow-tie antenna.

The modified printed bow-tie antenna presented exhibits a wide bandwidth (BW). The antenna consists of two identical printed bows, one on the top and one on the bottom of the substrate material. The top and bottom bows are connected to the microstrip feedline and the ground plane through a transition substrate, to match the bow-tie with the 50 Ω feedline, as illustrated in Figure 3.15.

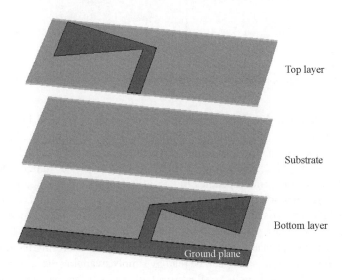

Figure 3.15 Bow-tie antenna geometry on a dielectric substrate. (Reproduced by permission of © 2005 IEEE [26])

3.5 Reflector Antennas

A periodic array of patches, printed on a dielectric substrate with a ground plane can be used as a planar reflector as shown in Figure 3.16. With a plane wave incident on the broadside, the power is reflected completely. The phase angle is determined by the patch length and patch width [27]. The reflection behaviour of this arrangement can be calculated using a spectral-domain method [28]. The phase angle varies over nearly 360°; thus, such elements can be used as reflection phase shifters. The phase angles calculated from a periodic structure can even be used for the design of reflecting elements in planar reflector antennas, as with the patches on a periodic grid, but with varying dimensions [29].

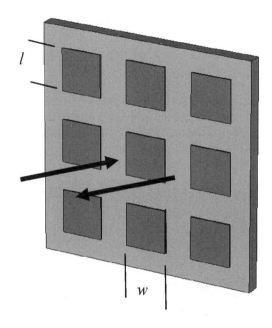

Figure 3.16   A periodic array of patches. (Reproduced by permission of © 2002 IEEE [27])

It should be noted that the phase angle of a reflection can be tuned by the sizes of the patches themselves, these mostly being far from resonance. Consequently, the reflect arrays exhibit quite low losses compared to a half-wavelength patch array, though with additional transmission lines for phase adjustment [30]. By making use of an independent choice of lengths and widths of the printed patches, different properties for the two polarisations can be realized, i.e. dual-function or dual-frequency antennas [29, 31]. The focusing array can be modified to include a polarisation twisting of the electromagnetic field, which, together with a printed polarising grid or a slot array, leads to a folded reflector antenna [32, 33].

In Figure 3.16, the patches are arranged on a quadratic grid. As a result, the reflection phase angle for the orthogonal polarisation can be read by simply swapping the length and width of a patch. The phase angle for an E-field parallel to the longer axis l of the element is approximately 70°, and the reflection phase angle is about $-110°$ for the orthogonal polarisation, i.e. a phase difference of 180° between the two polarisations.

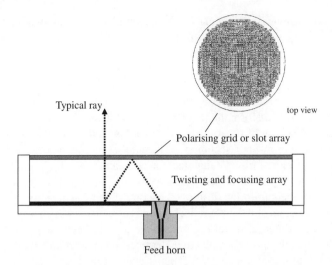

Figure 3.17 Basic principle of the folded reflector antenna and its top view. (Reproduced by permission of © 2002 IEEE [27])

The configuration of a printed folded reflector antenna is shown in Figure 3.17. The antenna consists of a feed, a planar polarisation filter and a printed *reflect array*. The feed can be either a cylindrical horn or a planar feeding line. The polarisation filter may be a grid or a resonant slot array, printed on a dielectric substrate and acting, at the same time, as a radome. The polarisation filter can reflect one selected polarisation and be transparent for the other polarisation.

The radiation from the feed is polarised so that it is reflected by a printed grid or slot array at the front of the antenna. Then the wave will be incident on the reflect array of printed patches. The patch axes of this array are tilted by $\pi/4$ with respect to the incident electric field. The electric field vector can be decomposed into the two components parallel to the two patch axes (Figure 3.18) and consequently, the reflection properties can be determined

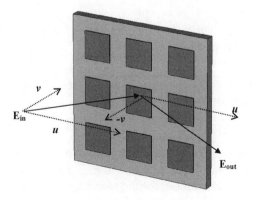

Figure 3.18 Vector decomposition of the incident and reflected electric fields for π of the reflection phase angle difference in v [27]

separately. The dimensions of the patches are selected in such a way that a phase difference of π occurs between the phase angles of these two reflected components. Superposition of the resultant field components then leads to a twisting of the polarisation by $\pi/2$ (Figure 3.18). The necessary π phase angle difference between the two field components of the reflected wave, can be achieved for a large number of combinations of length and width of the patches, if they differ by their absolute reflection phase angle. This degree of freedom is now used to adjust the required phase angles to transform the incident spherical wave into an outgoing plane wave.

In order to reduce the size of the folded reflector antenna, the effective focal length is kept short, generally resulting in poor scanning performance. But, as is known from the design of lens or reflector antennas, bifocal antennas for wide-angle scanning are possible. In such antennas, the single focal point is replaced by a focal ring [34, 35]. This requires, however, an additional degree of freedom in the design – selecting specific shapes for both surfaces of a lens [34], or both reflectors of a double reflector configuration [35]. As has been demonstrated in Reference [36], a printed grid can be used as the ground plane of a reflect array, and in combination with narrow dipoles as reflecting elements, this reflect array is nearly transparent for a wave in the orthogonal polarisation. This structure can therefore be used as a second reflector in a folded reflector antenna, replacing the polarising grid (Figure 3.19, top layer). Thus, it is appropriate to apply the principle of a bifocal antenna to a folded reflector antenna.

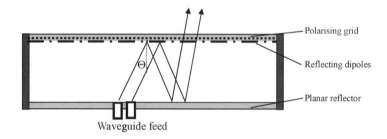

Figure 3.19   Cross-section of a dual-feed bifocal folded reflector antenna [27]

Similar to ray-tracing geometric optics [37, 38], the principle of a bifocal folded reflector antenna has the following features and assumptions:

- Higher-order effects such as diffraction or surface waves are not considered.
- There is no amplitude taper from the feed structure.
- Sidelobe suppression can be achieved at wide angles.

For the design of a conventional reflect array in Figure 3.17, the reflection phase angle is required at the point of incidence of a ray. In the case of a bifocal antenna, however, the relation between reflector properties of the reflect array and angles of incident/reflected rays should be re-evaluated. This can be explained by Figure 3.20. Two parallel rays with a small separation δr are incident on a planar structure at an angle Θ_1. They are reflected by the array structure and encounter a delay described by electrical lengths Φ_1 and Φ_2, and

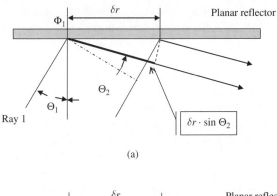

(a)

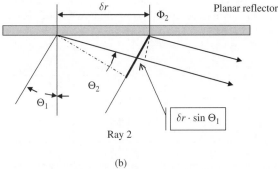

(b)

Figure 3.20 Principle of determining the relations between angles of incidence and reflection, and planar reflector properties: (a) ray 1 and (b) ray 2. (Reproduced by permission of © 2002 European Microwave Association (EuMA) [39])

leave the structure (approximately parallel) at an angle Θ_2. This requires the same path lengths for both rays. So that:

$$\delta r \cdot \sin \Theta_2 + \Psi_1 = \delta r \cdot \sin \theta_1 + \Psi_2 \tag{3.8}$$

With $\delta r \rightarrow 0$, this results in:

$$\frac{\delta \Psi}{\delta r} = \sin \Theta_2 - \sin \Theta_1 \tag{3.9}$$

Equation (3.9) can be used with a ray tracing procedure based on the defined feed position, beam angle and the symmetry of the antenna. The reflection phase angles for the two reflectors are then transferred into the respective patch dimensions. The front reflector is realised as a double-layer structure to improve stability; at the same time, the front substrate works as a radome (Figure 3.19, upper layer).

The lower substrate is a reflecting array which causes a twisting action. In a receiver, when a received signal goes through the top substrate, the ray is incident on the lower reflector. Following the law of reflection, the wave is reflected at the bottom substrate at the same angle while the wave polarisation is twisted by $\pi/2$. On the upper substrate, the received ray has to be reflected again to the feed point. From the known angles of incidence and reflection, the reflector properties ($\delta \Phi / \delta r$) can be computed at the reflection point using Equation (3.9).

Due to the symmetric structure, these properties must be the same on the opposite side of the antenna. In a transmitter, the reflection of a transmitting ray starting at the feed point can be traced up and down between the reflectors. Knowing the angle of incidence and the angle of the outgoing ray, the properties of the lower reflector can be determined at this point and, consequently, at the symmetric point. The whole procedure is continued until the edge of one of the reflectors is reached. A second set of data can be derived in the same way starting with a ray from the feed point to the centre of the upper reflector.

When several feeds are used, the centre feed opening can be designed to sit in the reflector plane, with the other ones protruding out of this plane for the best performance. Figure 3.21 shows E- and H-plane radiation characteristics at 76.5 GHz for the central beam (top) as well as the E-plane beams for the seven feeds (bottom). All beams are normalised to the power level of 0 dB. Beamwidths are between $3°$ and $3.3°$, the scanning range is $\pm 13.5°$ and the sidelobe level is better than -18 dB. The pattern is similar over a bandwidth of at least 76.5 ± 1 GHz [39].

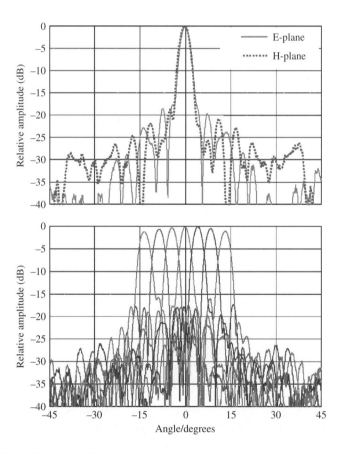

Figure 3.21 E- and H-plane radiation diagrams of central beam (top) and E-plane diagram for the seven feeds (bottom) of an antenna with 90 mm diameter and $\pm 13.5°$ scanning range. (Reproduced by permission of © 2002 European Microwave Association (EuMA) [39])

This antenna could be applied to smart antennas in two different ways. The first one resembles the conventional principle: a switching network may connect either of the feeds to a single radar front end. Low-loss microelectromechanical system (MEMS) switches could be applied to this network [39]. The other approach consists of connecting separate receivers to each feed to allow a parallel processing of all channels. The transmitter then must illuminate the complete detection range, which typically is done using a separate transmit antenna [40, 41].

Multifrequency systems can be constructed using planar technology. For example, both GSM (global system for mobile communications) [42] and 60 GHz antennas can be combined in one antenna structure. GSM works in the lower GHz range (i.e. 0.9 GHz, 1.8 GHz), while tie lines to the base stations are often realized in the millimetre wave range. This section presents a possible combination of antennas for the 900 MHz and the 60 GHz range in a common aperture. This makes it possible to have a very compact realisation of dual frequency systems. It can be applied to a small base station mounted on a wall of a building in a densely populated urban scenario. A reduced elevation beamwidth of the 900 MHz antenna can easily be achieved by placing further antenna elements (without integrating it with a millimetre wave antenna) below or above the antenna configuration described here.

A 60 GHz antenna can be implemented together with a 900 MHz one. The 900 MHz antenna is designed as a microstrip patch antenna over a simple metal box, while the millimetre wave antenna is integrated with this lower frequency antenna in the form of a folded reflector antenna [44, 45]. The configuration of this antenna is shown in Figure 3.22, together with a typical "ray" of the high-gain 60 GHz antenna.

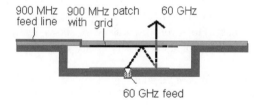

Figure 3.22 Setup of the dual frequency antenna [43]

The 900 MHz patch is designed on an inverted substrate. The substrate also works as a radome. The substrate is placed on top of a resonator box with air as the dielectric. When the current on the patch is concentrated mostly at the edges, a grid structure can easily be incorporated into the patch metallisation, acting as a polarising grid for the millimetre wave antenna. The electric field of the feed radiation is polarised and then reflected by the printed grid integrated into the low-frequency patch. Following this, the wave is incident on the lower substrate which has an array of printed dipoles (Figure 3.17, top right). The dipoles are tilted by 45° with respect to the incident electric field. The field can be decomposed into components parallel to the axes of the dipoles. The geometrical dimensions of the dipoles are designed in such a way that, a phase difference of 180° occurs between the two components of the reflected wave; leading to a twisting of the polarisation of the reflected wave by 90°. This type of twisting performance can be achieved by having a large number of combinations of length and width dipoles, which differ only by the absolute reflection phase angle. This overall phase shift is adjusted according to the focusing requirements. The original concept of this antenna is based on periodic structures using spectral domain calculations.

Such multiband antennas cover the mobile communication frequency band around 900 MHz and the communication band at 60 GHz, together with the ISM band around 61 GHz. The 900 MHz antenna is based on a resonator-backed microstrip patch antenna, while the millimetre wave antenna consists of a folded reflector antenna with a polarising grid integrated into the antenna patch; and a twisting and focusing planar reflector placed on the bottom of the 900 MHz antenna box. This configuration could be an antenna solution for GSM and 60 GHz access points.

3.6 Millimetre Wave Design Considerations

In millimetre waves, the thickness of the dielectric between the patch or patch stack and the ground plane has a strong influence on the bandwidth of the antenna [46] and, consistently, the simulations and measurements show a wider impedance bandwidth for the circular single-element (Figure 3.3 (a)) than the stacked two-element patch antenna (Figure 3.7 (c)) [47].

The diameter of the patch can be designed to be 890 μm with the silver paste spreading causing an extra 60 μm in patch diameter. This explains why the observed centre frequency spots are sometimes detuned below 60 GHz. The 25-μm-shifted feed point position corrects, to some degree, the effect of the increased realised patch size.

The spreading of the silver paste during the firing stage of the low-temperature co-fired ceramics (LTCC) (see Chapter 9) [48] processing, has occurred quite often with the Ferro A6-S materials [49]. It is difficult to predict how much the spreading will be compensated by the layout design in each case. However, research work shows that gold paste of the Ferro A6-S system does not practically spread at all, and hence it is the preferred choice for these types of applications.

The transition can be designed using a coupling slot with the coplanar waveguide line, as shown in Figure 3.23. In general, a slot transition cannot be used if wideband functioning is desired, but in this particular case this is not regarded as a problem. The design of the slot transition is quite simple. The width of the slot is selected so that its realisation in the LTCC process is feasible. For this implementation 150 μm is considered reasonable. The length of the slot from end to end is close to the corresponding electrical half-wavelength in the dielectric medium used, which is about 1.0 mm in this case. Both the coplanar waveguide and SL continue slightly over the slot region, and hence form two stubs in the transition. The exact dimensions of the slot and the stubs can be calculated using a simulator, to give the desired trade-off between adequate return loss and band width. For this transition, the simulated return loss is about −39 dB and the insertion loss is close to −0.62 dB at 60 GHz. The impedance matching bandwidth for the return loss of at least −15 dB is in the range of 54–66 GHz.

The minimum insertion loss for a single transition seems to be 1.1 dB, though the S_{11} and S_{22} may have different frequency responses if there is a variation in the physical dimensions compared to the realised conductor patterns; which will cause some asymmetry [50]. Despite the deteriorated performance caused by the dimensional tolerances, the functioning of the transition can be regarded as adequate to achieve the aimed for antenna radiation gain measurements.

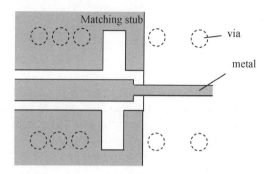

Figure 3.23 Coplanar waveguide-to-stripline transition

3.7 Production and Manufacture

Millimetre wave production allows small tolerances and low loss. This section introduces several manufacturing technologies that are suitable for millimetre wave antenna production in terms of reducing loss and increasing printing accuracy.

3.7.1 Fine Line Printing

The fine line technique is based on print-and-etch techniques, using a standard dry film etch-resist [51]. This allows large areas of very fine printed circuit boards (PCBs) to be manufactured on a range of laminates with higher reliability and at a lower cost than alternative techniques. Tracks and gaps can be fabricated on various laminates, including standard FR4, FR5 and Flex [52] as well as soft boards (e.g. Teflon™) and ceramic substrates.

The fine line technique has particular application to higher-density microelectronic packaging for portable mobile electronics, particularly for telecoms applications, where functionality is at a premium. Among the key areas where this interconnect technology will help is in "flip chip" assembly, which is key for many applications where high-speed signal performance and packaging are demanded. This is important in mobile telecom systems.

The new technology allows 25 μm tracks and gaps to be produced reliably and at a lower cost than current techniques do. It extends printing circuit technology into the millimetre wave range and addresses the needs of a wide range of microelectronic applications, from laptops to medical instruments.

The move towards very small devices such as chip scale packages and flip chip assemblies is constrained by the density limitations of conventional PCB technology. In most imaging processes used in PCB technology, it is difficult to go below a 100 μm resolution. One of the alternatives to conventional imaging is the new technology of laser structuring. It creates a structure in resist (tin or organic coating) by direct laser ablation. After ablating the resist, the structure is etched and the remaining resist is stripped. The direct patterning of tin and organic coatings has been studied to create masks for the etching process. The target has been to achieve lines/spaces below 125 μm [53].

3.7.2 Thick Film

Millimetre wave circuitry requires high-resolution etching technology, and its functional density is increasing. It is common to use ceramic materials, which include thick-film conductors on aluminates and circuits using Green Tape™ low-temperature co-fired ceramics [48]. Conductor patterning techniques include conventional screen printing, Fodel® photoimaging and etching processes. Thin-film depositions have also been considered since this has been the most viable technology from which to fabricate millimetre wave circuits [51].

It is important to note that the fabrication of millimetre wave circuits requires close collaboration with the designer, from circuit layout to final productions, because of the special geometries required for: transmit and receive lines, for waveguides, to limit reflection and propagation losses, as well as for the construction of specific components, such as antennas, couplers and dividers.

Since the mid-1990s, more opportunities have arisen for the use of ceramic circuitries in high-frequency modules; with more information available on their performance in the millimetre wave range and new materials on offer, including LTCC and FODEL® photodefinable conductors [54].

3.7.3 Thin Film

Thin-film technology has been the traditional method used to manufacture microwave circuitry for many years. With outstanding line resolution, excellent conductor edge definition, combined with superb ceramic substrate properties at high frequency and stable thermal behaviour; which are all ideal attributes of circuitry for use in portable devices. However, this approach has disadvantages in terms of cost. Complex modules are often assembled in a special hermetic housing using a patchwork arrangement of certain substrates. The major reason used to divide the circuit into subcircuits is related to the yield figures obtained on large substrates. Positioning accuracy of these substrates is crucial to avoid gaps and related impedance changes. The housing itself needs to have expensive hermetic RF interconnections. Multilayer substrates based on LTCC [48] offer a variety of options for microwave designs. DC connections and digital control functions can be implemented in separate layers, chip tailored cavities can be used to improve the return loss of the signal interconnections, various transmission line types as well as waveguides are available; and the hermetic substrate itself can be used as part of the package with integrated feedthroughs. Embedded resistors and capacitors are additional features which will further shrink the size of designs. However, fine line printing resolution and associated tolerances may well be the restricting factors in minimizing design size. Although lines and spaces down to $50\,\mu m$ are achievable, this is not sufficient for certain elements like edge coupled filters, couplers, etc.

A recent approach, FINEBRID, combining the advantages of both technologies, was developed and evaluated within a funded program [51]. Thin-film structures were applied on fired LTCC substrates without special surface treatment. This process allows a combination of printed thick-film and thin-film structures on the surface. Hence, the combined technology offers improved technology features such as smaller lines and spaces. Thin-film features can be reduced to the necessary areas, and special thick-film materials for hermetic sealing can be applied, thus providing options to reduce cost, size and weight.

3.7.4 System-on-Chip

Increasing demand for low-cost, broadband, high-speed and small wireless communication devices, especially in the millimetre wave frequency range, has turned the SoC (system-on-chip) solution into an important technique to satisfy these demands [55]. One of the most important problems in the performance of on-chip antennas in the millimetre wave range is substrate losses. By using micromachining techniques [56], it is possible to remove unwanted regions of the substrate and thus reduce substrate losses.

In the example of silicon technology, (100) silicon substrates with a thickness of $550 \mu m$ are typically used for the realisation of on-chip antennas [56]. The fabrication of devices requires two steps of back- and front-side processing. The back-side etching is performed in a KOH (potassium hydroxide) solution of a concentration of 8 moles, and at a temperature of $52\text{--}58\,°C$. During this step, silicon is removed through the openings in the masking layer for a period of 25–30 hours. Since the etching step is rather extended, it is important that the masking layer can withstand long exposures to etching chemicals. Thus it can be seen that there are a number of fabrication technologies available. Each has its advantages and drawbacks. However, the final application will determine the choice of process.

References

[1] N. G. Alexopoulos and I. E. Rana, 'Mutual Impedance Computation between Printed Dipoles', *IEEE Transactions on Antennas and Propagation*, **AP-29**, January 1981, 106–111.

[2] R. S. Elliot and G. I. Stem, 'The Design of Microstrip Dipole Arrays Including Mutual Coupling, Part 1: Theory; Part 11: Experiment', *IEEE Transactions on Antennas and Propagation*, **AP-29**, September 1981, 757–765.

[3] D. Pozar, 'Considerations for Millimetre Wave Printed Antennas', *IEEE Transactions on Antennas and Propagation*, **31**(5), September 1983, 740–747.

[4] K. R. Carver and J. W. Mink, 'Microstrip Antenna Technology', *IEEE Transactions on Antennas and Propagation*, **AP-29**, January 1981, 2–24.

[5] R. F. Harrington, *'Time-Harmonic Electromagnetic Fields'*, McGraw-Hill, New York, 1961.

[6] Rogers Corporation website, http://www.**rogers**corporation.com/

[7] DuPont Microcircuit Material website, http://www.**dupont**.com/mcm/

[8] Dielectric Chart, Emerson and Cuming Microwave Products, Massachusetts.

[9] K. S. Yngvesson, T. L. Korzeniowski, R. H. Mathews, P. T. Parrish and T. C. L. G. Sollner, 'Plane Millimetre Wave Antennas with Application to Monolithic Receivers', *Proceedings of SPIE*, **337** (Milli-meter Wave Technology), 1982.

[10] M. Irsadi Aksun, Shun-lien Chuang *et al.*, 'On Slot-Coupled Microstrip Antennas and Their Applications to CP Operation – Theory and Experiment', *IEEE Transactions on Antennas and Propagation*, **38**(8), August 1990, 1224–1230.

[11] O. Lafond, M. Himdi and J. P. Daniel, 'Aperture Coupled Microstrip Patch Antenna with a Thick Ground Plane in Millimetre Waves', *IEE Electronics Letters*, **35**(17), August 1999, 1394–1396.

[12] W. S. T. Rowe and R. B. Waterhouse, 'Broadband Coplanar-Waveguide Fed Stacked Patch Antenna', *Electronics Letters*, **35**(9), 1999, 681–682.

[13] David Pozar, 'Microstrip Antennas', *Proceedings of IEEE*, **80**(1), January 1992, 79–91.

[14] Wikipedia,'Babinet's_Principle', http://en.wikipedia.org/wiki/Babinet's_principle.

[15] Gerald Oberschmidt, Veselin Brankovic and Dragan Krupezevic, 'V-Slot Antenna for Circular Polarization', US Patent 2002000943.

[16] K. S. Yngvesson, D. H. Schaubert, T. L. Korzeniowski, E. L. Kollberg, T. Thungren and J. F. Johansson, 'Endfire Tapered Slot Antennas on Dielectric Substrates', *IEEE Transactions on Antennas and Propagation*, **AP-33**, December 1985, 1392–1400.

[17] K. S. Yngvesson *et al.*, 'The Tapered Slot Antenna-A New Integrated Element for Millimetre-Wave Applications', *IEEE Transactions on Microwave Theory and Technology*, **37**, February 1989, 365–374.

[18] Felix K. Schwering,'Millimetre Wave Antennas', *Proceedings of the IEEE,* **80**(1), January 1992, 92–102.

[19] William R. Deal, Noriaki Kaneda (Student), James Sor, Yongxi Qian and Tatsuo Itoh, 'A New Quasi-Yagi Antenna for Planar Active Antenna Arrays', *IEEE Transactions on Microwave Theory and Techniques*, **48**(6), June 2000, 910–918.

[20] Y. Qian and T. Itoh, 'A Broadband Uniplanar Microstrip-to-CPS Transition', *Asia–Pacific Microwave Conference Digest*, December 1997, 609–612.

[21] N. G. Alexopoulos, P. B. Ketehi and D. B. Rutledge, 'Substrate optimization for integrated circuit antennas', *IEEE Transactions on Microwave Theory and Techniques*, **MTT-31**, July 1983, 550–557.

[22] I. Oppermann, M. Hämäläinen and J. Iinatti, *'UWB: Theory and Applications'*, John Wiley & Sons, Ltd, Chichester, September 2004.

[23] R. N. Simons, *'Coplanar Waveguide Circuits, Components, and Systems'*, John Wiley & Sons, Inc., New York, 2001.

[24] Hojr Sedaghat-Pisheh1, Mahmoud Shahabadi 2 and Shamsodin Mohajerzadeh 1, 'Design, Simulation, and Fabrication of an On-Chip Antenna Fabricated Using Silicon Micromachining for Broad-Band Millimetre-Wave Wireless Communications', Proceedings of the 36th European Microwave Conference, 2006.

[25] V. K. Varadan, K. J. Vinoy and K. A. Jose, *'RF MEMS and Their Applications'*, John Wiley and Sons, Ltd, Chichester, December 2002.

[26] Abdelnasser A. Eldek, Atef Z. Elsherbeni, 'A Microstrip-Fed Modified Printed Bow-Tie Antenna for Simultaneous Operation in the C and X-Bands', IEEE International Radar Conference, May 2005, pp. 939–943.

[27] W. Menzel, D. Pilz and M. Al-Tikriti, 'Millimetre-Wave Folded Reflector Antennas with High Gain, Low Loss, and Low Profile', *IEEE Antennas and Propagation Magazine*, **44**(3), June 2002, 24–29.

[28] R. Mittra *et al.*, 'Techniques for Analyzing Frequency Selective Surfaces – A Review', *Proceedings of the IEEE*, **76**(12), December 1988, 1593–1615.

[29] D. M. Pozar, S. D. Targonski and H. D. Syrigos, 'Design of Millimetre Wave Microstrip Reflectarrays', *IEEE Transactions on Antennas and Propagation*, **AP-45**, 1997, 287–296.

[30] R. D. Javor, X.-D. Wu and K. Chang, 'Design and Performance of a Microstrip Flat Reflectarray Antenna', *Microwave and Optical Technology Letters*, **7**(7), 1994, 322–324.

[31] D. Pilz and W. Menzel, 'Periodic and Quasi-Periodic Structures for Antenna Applications', Proceedings of the 29th European Microwave Conference, Vol. III, Munich, Germany, 1999, pp. 311–314.

[32] W. Menzel, D. Pilz and R. Leberer, 'A 77 GHz FM/CW Radar Front-End with a Low-Profile, Low-Loss Printed Antenna', *IEEE Transactions on Microwave Theory and Techniques*, **MTT-47**(12), December 1999, 2237–2241.

[33] W. Menzel and D. Pilz, 'Printed Quasi-Optical Wave Antennas', Millennium Conference on *'Antennas and Propagation'*, AP2000, Davos, Switzerland, 2000, Session 3A2-1, Paper 0023.

[34] P. S. Holt and A. Mayer, 'A Design Procedure for Dielectric Microwave Lenses of Large Aperture Ratio and Large Scanning Angle', *IRE Transactions on Antennas and Propagation*, **AP-5**, 1957, 25–30.

[35] A. Y. Niazi and P. J. Mitchell, 'Millimetre Wave Phase Corrected Reflector Antenna', IEE International Conference on *'Antennas and Propagation'*, ICAP'83, Norwich, England, April 1983, Part 1, pp. 51–54.

[36] W. Menzel and D. Pilz, 'Printed mm-Wave Folded Reflector Antennas withHigh Gain, Low Loss, and Low Profile', IEEE Antennas and Propagation Conference, Salt Lake City, Utah, July 2000, Vol. 2, pp. 790–793.

[37] P. S. Holt and A. Mayer, 'A Design Procedure for Dielectric Microwave Lenses of Large Aperture Ratio and Large Scanning Angle', *IRE Transactions on Antennas and Propagation*, **AP-5**, 1957, 25–30.

[38] A. Y. Niazi and P. J. Mitchell, 'Millimetre Wave Phase Corrected Reflector Antenna, IEE International Conference on *'Antennas and Propagation'*, ICAP'83, Norwich, England, April 1983, Part 1, pp. 51–54.

[39] W. Menzel, M. Al-Tikriti and R. Leberer, 'A 76 GHz Multiple-Beam Planar Reflector Antenna', 34th European Microwave Conference, 2002, pp. 1–4.

[40] C. Metz, E. Lissel and A. F. Jacob, 'Planar Multiresolutional Antenna for Automotive Radar, 31st European Microwave Conference, London, 2001, pp. 335–338.

[41] M. Younis, A. Herschlein, Y. J. Park and W. Wiesbeck, 'A Parallel-Plate Luneburg Lens Sensor Concept for Automatic Cruise Control Applications, 31st European Microwave Conference, London, 2001, pp. 339–342.

[42] Wikipedia, 'GSM', http://www.wikipedia.org/

[43] Wolfgang Menzel, Maysoun Al-Tikriti and Maria Belen Espadas Lopez, 'A Common Aperture, Dual Frequency Printed Antenna (900 MHz and 60 GHz)', *Electronics Letters*, **37**(17), 16 August 2001, 1059–1060.

[44] D. Pilz and W. Menzel, 'Folded Reflectarray Antenna', *Electronics Letters*, **34**(9), April 1998, 832–833.

[45] D. Pilz and W. Menzel, 'Printed Millimetre-Wave Reflectarrays' *Annales de Telecommunications*, **56**(1–2), 2001, 2–11.

[46] K. R. Carver and J. W. Mink, 'Microstrip Antenna Technology', *IEEE Transactions on Antennas and Propagation*, **29**, January 1981, 18–20.

[47] Antti Vimpari, Antti Lamminen and Jussi Säily, 'Design and Measurements of 60 GHz Probe-Fed Patch Antennas on Low-Temperature Co-fired Ceramic Substrates', Proceedings of the 36th European Microwave Conference, 2006, pp. 854–857.

[48] Wikipedia, 'Low Temperature Co-fired Ceramics', http://www.wikipedia.org/

[49] Ferro, 'LTCC A6 System for Wireless Solutions', Ferro® Electronic Material.

[50] Antti Vimpari, Antti Lamminen (2) and Jussi Säily, 'Design and Measurements of 60 GHz Probe-Fed Patch Antennas on Low-Temperature Co-fired Ceramic Substrates', Proceedings of the 36th European Microwave Conference, 2006.

[51] G. Vanrietvelde, E. Polzer, S. Nicotra, J. Mueller and A. Brokmeier, 'Microwave and Millimeter Wave Applications: A New Challenge for Ceramic Thick Film Technology', IEE Seminar on *'Microwave Thick Film Materials and Circuits'*, 2007, paper 2002/097, pp. 6–10.

[52] R. S. Khandpur, *'Printed Circuit Boards'*, McGraw-Hill Professional, September 2005.

[53] W. Falinski, G. Koziol and J. Borecki, 'Laser Structuring of Fine Line Printed Circuit Boards', 28th International Spring Seminar on *'Electronics Technology: Meeting the Challenges of Electronics Technology Progress'*, 19–20 May 2005, pp. 196–201.

[54] DuPont Electronic Material Data Sheet, 'Fodel® 6778 Conductor', http://www.dupont.com/

[55] Y. P. Zhang, 'Recent Advances in Integration of Antenna on Silicon Chip and Ceramic Package', IEEE International Workshop on *'Antenna Technology'*, 2005, pp. 151–154.

[56] Gildas P. Gauthier, Linda P. Katehi and Gabriel A. I. Rebeiz, 'A 94 GHz Aperture-Coupled Micromachined Microstrip Antenna', *IEEE MTT-S Digest*, 1998, 993–996.

4

Horn Antennas

Attention is now turned to three-dimensional structures; and the basic features of three-dimensional waveguide launchers and their applicability to millimetre wave systems. An outline of the work presented in this chapter is as follows. In Section 4.1, waveguide modes in smooth walled horns are reviewed and circular cross section multimode horn designs are discussed in Section 4.2. The careful control of waveguide modes is shown to be essential in maintaining both the aperture distribution in the horn and also its polarization response. A brief mention of corrugated horns then follows. Then, quasi-integrated horn designs are presented in Section 4.3 and the corresponding radiation patterns are shown. Cylindrical horns are presented in Section 4.4 along with a description of the design procedure. Subsequently, in Sections 4.5 and 4.6, two novel configurations, the tilt horn and the dielectric sectoral horn, for millimetre wave design are discussed.

4.1 Waveguide Modes

First of all, it is useful to review smooth walled waveguide modes before moving on to antenna design. Figure 4.1 shows three types of smooth walled rectangular horns with a rectangular waveguide feed.

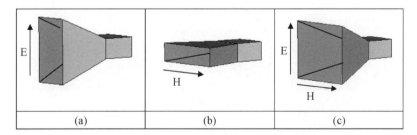

(a)	(b)	(c)

Figure 4.1   (a) E-plane, (b) H-plane and (c) pyramidal horns

The dominant propagation mode for a rectangular waveguide is the TE_{10} mode, which has the lowest cut-off frequency. Also, it is dependent only on the length of the longest side of the

---

Millimetre Wave Antennas for Gigabit Wireless Communications   Kao-Cheng Huang and David J. Edwards
© 2008 John Wiley & Sons, Ltd

waveguide. If multiple modes can propagate simultaneously without proper control, this may cause unexpected amounts of dispersion, distortion and variable operation.

In Figure 4.2, the cut-off frequency for TE_{10} can be expressed as $f_{10} = c/2a$. For a higher mode, $f_{20} = c/a$ and $f_{01} = c/2b$. If $b < a/2$, f_{01} is larger than f_{20}. If $a/2 < b < a$, f_{01} is smaller than f_{20}. In order to achieve the widest possible usable bandwidth for the TE_{10} mode, the guide dimensions must satisfy $b < a/2$, so that the bandwidth is the interval between f_c and $2f_c$.

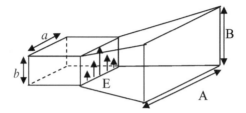

Figure 4.2 TE_{10} mode electric field over a waveguide aperture

The gain of the horn antenna can be expressed as:

$$G = 4\pi \frac{U_{max}}{P_{rad}} = e \frac{4\pi}{\lambda^2} AB$$

where:

U_{max} is the maximum intensity
P_{rad} is the total power transmitted through the aperture
e is aperture efficiency
AB is the area of the horn aperture as shown in Figure 4.2

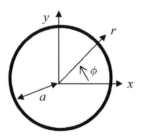

Figure 4.3   A circular waveguide of radius a

In a circular waveguide, there are a number of solutions to the wave equation that satisfy the boundary conditions. These solutions are based on Bessel functions determined by the radius of the waveguide and the frequency and wavelength in the guide. For a given frequency, the modes that can be supported are determined by the waveguide radius.

Circular waveguides offer manufacturing advantages over rectangular waveguides in that production is much simpler because only one dimension – the radius – needs to be considered. Calculation for a circular waveguide requires the application of Bessel functions. Basic formulas of the modes that can be supported in a circular waveguide are listed in Table 4.1 [1].

Table 4.1 Formulas of circular waveguide modes

Variable	TE Modes (see Note 1)	TM Modes (see Note 2)
H_z	$J_n\left(\frac{p'_{nm}r}{a}\right)e^{-j\beta_{nm}z}\begin{cases}\cos(n\phi)\\\sin(n\phi)\end{cases}$	0
E_z	0	$J_n\left(\frac{p_{nm}r}{a}\right)e^{-j\beta_{nm}z}\begin{cases}\cos(n\phi)\\\sin(n\phi)\end{cases}$
H_r	$-\frac{j\beta_{nm}p'_{nm}}{ak_{c,nm}^2}J'_n\left(\frac{p'_{nm}r}{a}\right)e^{-j\beta_{nm}z}\begin{cases}\cos(n\phi)\\\sin(n\phi)\end{cases}$	$-\frac{E_\phi}{Z_{e,nm}}$
H_ϕ	$-\frac{jnp'_{nm}}{rk_{c,nm}^2}J_n\left(\frac{p'_{nm}r}{a}\right)e^{-j\beta_{nm}z}\begin{cases}-\sin(n\phi)\\\cos(n\phi)\end{cases}$	$\frac{E_r}{Z_{e,nm}}$
E_r	$Z_{h,nm}H_\phi$	$-\frac{j\beta_{nm}p_{nm}}{ak_{c,nm}^2}J'_n\left(\frac{p_{nm}r}{a}\right)e^{-j\beta_{nm}z}\begin{cases}\cos(n\phi)\\\sin(n\phi)\end{cases}$
E_ϕ	$-Z_{h,nm}H_\rho$	$-\frac{jn\beta_{nm}}{rk_{c,nm}^2}J_n\left(\frac{p_{nm}r}{a}\right)e^{-j\beta_{nm}z}\begin{cases}-\sin(n\phi)\\\cos(n\phi)\end{cases}$
β_{nm}	$\left[k_0^2-\left(\frac{p'_{nm}}{a}\right)^2\right]^{1/2}$	$\left[k_0^2-\left(\frac{p_{nm}}{a}\right)^2\right]^{1/2}$
$k_{c,nm}$	$\frac{p'_{nm}}{a}$	$\frac{p_{nm}}{a}$
$\lambda_{c,nm}$	$\frac{2\pi a}{p'_{nm}}$	$\frac{2\pi a}{p_{nm}}$

Note 1. The values of p'_{nm} for TE Modes can be shown in the following table:

n	p'_{n1}	p'_{n2}	p'_{n3}
0	3.832	7.016	10.174
1	1.841	5.331	8.536
2	3.054	6.706	9.970

Note 2. The values of p_{nm} for TM Modes can be shown in the following table:

n	p_{n1}	p_{n2}	p_{n3}
0	2.405	5.520	8.654
1	3.832	7.016	10.174
2	5.135	8.417	11.620

The dominant mode of a circular waveguide is the TE_{11} mode, which has the lowest cut-off frequency, and is the one normally used. The ratio of the cut-off frequency between the higher modes and the TE_{11} mode is drawn in Figure 4.4. A sketch of the field lines in the transverse plane for this mode is shown in Figure 4.5.

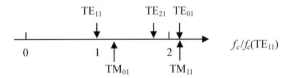

Figure 4.4 The cut-off frequency ratio of the first few TE and TM modes to the dominant mode (TE_{11})

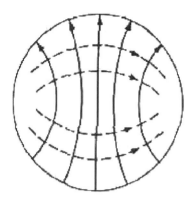

Figure 4.5   Field lines for the TE_{11} mode in circular waveguide

4.2 Multimode Horn Antennas

Circular aperture horns have been used as feed for many years in microwave antenna systems. The use of multimodes in a circular horn can result in sidelobe suppression in both the E- and H-planes, along with the additional benefit of beamwidth equalisation. It is the launching of several modes and the control of the aperture field that enables the device to deliver high performance, and the ability to balance the generated modes gives the designer a certain amount of freedom to optimise the aperture distribution. The structure of a multimode circular horn is simple and economical to fabricate, and contributes to low VSWR and can minimise dissipation loss. Therefore, the multimode circular horn with a circularly symmetric pattern, is a near-ideal feed for a low-noise Cassegrain, high-aperture efficiency, antenna in the millimetre wave and submillimetre wave bands.

This multimode horn has the following features:

1. An aperture distribution may be synthesised to realise an optimum radiation pattern.
2. Linear combinations of the radiation pattern functions result in low sidelobes and higher secondary gain, when used to illuminate a reflector antenna.
3. Performance is derived theoretically to match the aperture distribution of the horn to the required illumination function of the reflector aperture. This is in turn derived from the specified radiation performance of the antenna system.

The basic structure of the multimode horn contains elements within the throat of the horn, that allow or stimulate mode conversion as the wave propagates towards the aperture. These elements may be step or discrete discontinuities or indeed corrugations to maintain a given hybrid mode structure. The multimode circular horn can be realised in a number of forms, including single-flare-angle change, double-flare-angle change or step-discontinuity. Figure 4.6 shows a configuration of the double-flare angle change horn. The desired higher-order modes are generated at the flare-angle discontinuity.

The TE_{1m} and TM_{1n} modes may propagate in the oversized circular waveguide, because the diameter of the circular waveguide may be sufficiently large relative to the guide wavelength. If this diameter changes, then the boundary conditions need to be maintained and consequently, a different set of modes can be generated in the guide. If a basic TE_{11} mode is

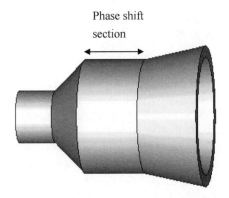

Phase shift
section

Figure 4.6 Multimode circular horn configuration

incident at a flare-angle discontinuity, the first-order forward modes are the TM_{11} and TE_{12} modes, and the second-order modes fitting the requirement to match the curved phase front of the TE_{11} mode are the TE_{13} and TM_{12} modes. The power conversion coefficients and phase of propagating modes, excited by a symmetric flare-angle change in the circular waveguide can be accurately computed by a modal analysis of the discontinuity [2] (in effect seeking to establish which set of modes summed together will satisfy the boundary conditions). The differential phase shift between the dominant mode TE_{11}, and higher-order modes TE_{12}, TE_{13}, TM_{11} and TM_{12} at the aperture of the horn may be adjusted to a certain degree by adjusting the length of the phase-shift section of the circular horn. This can have uses in constructing particular aperture distributions for special applications such as shaped-beam reflector antennas.

Mode generation can be a switched feature in certain cases where mechanical or electronic features in the throat of the horn convert modes or not, as required. An example utilizing TE_{21}–TM_{01} (see Figure 4.7) mode-switched feeds is demonstrated below. Its use as a tracking method will be introduced in Chapter 8. In general, attention should be concentrated on the phase-shift section of the flare-angle change horn. If the phase-shift section is considered

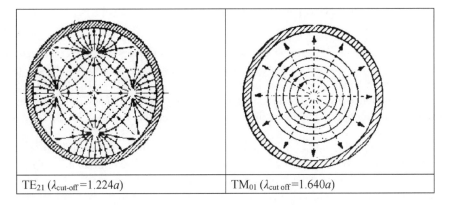

| TE_{21} ($\lambda_{\text{cut-off}}=1.224a$) | TM_{01} ($\lambda_{\text{cut off}}=1.640a$) |

Figure 4.7 Field distribution for TE_{21} and TM_{01} modes. *Source*: Baden Fuller, A. J. *Microwaves: an introduction to microwave theory and techniques*: 2nd Ed. Oxford; New York: Pergamon, 1979

carefully, the TE and TM modes at the aperture of the multimode horn can be designed to match a specific aperture distribution.

A mode conversion cavity, shown in Figure 4.8, can be designed to provide mode conversion. The device comprises a length of circular waveguide which is coupled to individual mode converters, e.g. frequency-tuned cavities [3], for the selected modes. In effect, the slot in the main waveguide imposes the boundary conditions which only the desired modal field pattern satisfies. Energy from this mode is coupled into the cavity and reflected back into the main waveguide by the end wall. Each individual mode converter contains a PIN diode. When the diode is not conducting the converter has little or no effect on the modal structure within the main waveguide (it is matched). The energy is in effect returned to the main guide in the same state as it was coupled out. When the converter is active, and the PIN diode conducting the slot and cavity are unmatched (the cavity end wall is in effect moved) and the mode is generated.

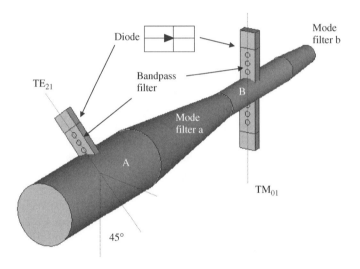

Figure 4.8   A perspective view of an example of a mode conversion module suitable for obtaining complete tracking information with cross-polar compensation from the TM_{01} and TE_{21} (V-polarised) higher-order modes with circularly polarised signals [3]

In many applications it is desirable to use the converters in pairs, i.e. two converters positioned diametrically opposite one another on the waveguide. In the example here a pair of TM_{01} generators are axially spaced and perpendicular to a pair of TE_{21} (H-polarised) generators. This arrangement converts received signals in only one plane of polarisation. Two planes of polarisation can be converted by providing four TM_{01} generators and four TE_{21} (H-polarised) generators.

The ability to generate and control these modes enables the aperture distribution of the horn to be switched. For the higher order modes discussed here, careful control of the phase and amplitude of the higher order modes relative to the fundamental mode can create a horn aperture distribution which has a phase tilt. Combining this horn aperture distribution with a reflector antenna enables the antenna beam to be directed. Alternatively if the mode generators are replaced by higher order mode couplers (devices which couple the energy contained in the higher order mode) a receiving antenna may be constructed which uses the relative phase and amplitude (with respect to the fundamental) of the higher order modes to detect the direction of arrival of the source signal. This is the basis of higher order mode monopulse tracking.

In such feeds, it is essential to employ a mode filter in a portion of waveguide that supports only the fundamental, as shown in Figure 4.8, between the mode conversion module and the receiver. In practice the conversion is not 100 % efficient and unconverted energy must be prevented from propagating down the waveguide in the reverse direction. A mode rejection filter provides this function.

In [3] in order to achieve the desired performance, it is important to maintain correct phase relationships at the launch aperture (i.e. at the end of the feed). The relationship is such that the higher-order mode is in phase quadrature with the fundamental (and mode converters are located so as to produce this relationship). Ideally, the amplitude is not affected by the interaction but the phase is tilted.

In a tracking feed horn, the diodes of the auxiliary waveguides are controlled so that each mode converter is rendered operative in turn while the others are inoperative, the converted fundamental mode created by the operative mode converter combines with the existing fundamental mode to produce a beam shift in an antenna system. The single frequency filtered output from the feed is then connected to a tracking receiver so that the amplitude of the signal as the beam is deflected can be used to calculate the pointing direction for the antenna.

For two axis beam deflection TE_{21} (vertically-polarised) TM_{01} operate sequentially, producing alternate shifts of the beam vertically and sideways. The vertically shifted beam will provide vertical/elevation plane tracking information, and the horizontally shifted beam will provide horizontal/azimuth plane tracking information. Figure 4.9 illustrates how the radiation patterns of the TE_{21} and TM_{01} modes combine to form the radiation pattern of the shifted beam in each case.

Figure 4.10 shows an alternative design for the mode converter of Figure 4.8. In this case there is only a single TM_{01} mode converting auxiliary waveguide, and an additional identical TE_{21} mode converter is coupled longitudinally to the first central waveguide section diametrically opposite the other auxiliary waveguide. Operation of the TM_{01} mode converter simultaneously with each of the TE_{21} mode converter alternately will produce alternate

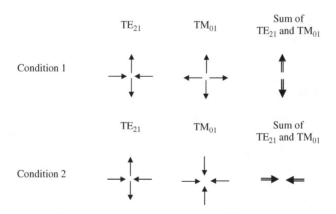

Figure 4.9 Electric field pattern diagrams illustrating how the higher-order modes in the module of Figure 4.8 combine to produce the cross-polar compensated tracking information

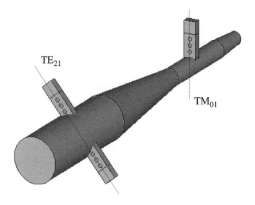

Figure 4.10   Alternative design for mode conversion module [3]

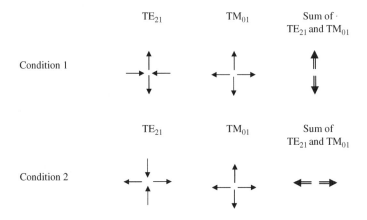

Figure 4.11   Electric field pattern diagrams illustrating how the higher-order modes in the module of Figure 4.10 combine to produce the cross-polar compensated tracking information

beam shifts, giving vertical/elevation plane tracking information and horizontal/azimuth plane tracking information. The electric field pattern diagrams can be seen in Figure 4.11.

4.3 Integrated Horn

The integrated horn antenna array was first developed in 1990 [4]. It consists of a dipole antennas suspended on a dielectric membrane inside a pyramidal cavity etched in silicon. It can also work with different planar antennas such as patches, as shown in Figure 4.12.

The power received by the horns excites the patch antenna. The dipole antenna, detectors and electronic circuitry are all integrated on the same side of the silicon wafer. The horn antennas generate excellent radiation patterns with directivities of between 10 and 13 dB. The aperture efficiency includes the intrinsic aperture loss due to the non-uniform field distribution at the aperture of the horn, the cross polarisation loss and the mismatch loss. The integrated horn

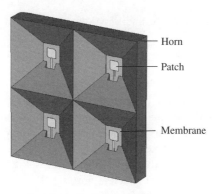

Figure 4.12   An integrated horn antenna array with a single polarisation

together with the planar antenna is an efficient antenna at millimetre wave frequencies, and could be seriously considered for application in gigabit wireless communications.

A dual-polarised antenna consists of a patch with two feeds perpendicular to each other, suspended on the same membrane inside the horn cavity (Figure 4.13) The feeds are coupled to an orthogonal set of patch modes and therefore are effectively isolated from each other.

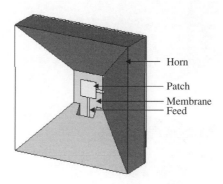

Figure 4.13   Dual-polarised patch-fed horn antenna

Conventional mechanically machined horn antennas integrated with waveguide cavities have been the standard at microwave and millimetre wave frequencies since they were first implemented during World War II. Very high antenna gain and essentially ideal (100 %) antenna aperture efficiency can be achieved using these structures. However, they are expensive, bulky and very difficult to incorporate into arrays. In order to overcome these issues, new developments using micromachining to fabricate the horn antenna structures have been developed [5]. In these structures, the active elements and their planar antennas are fabricated on a free-standing thin SiN membrane, which is suspended over a silicon pyramidal horn that is formed by anisotropic etching, or micromachining. The side walls of this micromachined structure can then be coated with gold to form a horn antenna. Compared to conventional waveguide horn antennas, this novel micromachined structure has several major advantages and it is an easier method to fabricate fine three-dimensional structures from than by using photolithography.

Using these methods, horn antennas with micrometre precision can be easily defined and inexpensively mass produced. They are fabricated on Si or GaAs wafers, a process that is compatible with thin-film technology. Thus, active millimetre wave and RF elements, such as amplifiers, mixers and detectors, local oscillators and post-detection signal processors, can be integrated monolithically with the antenna structures to form monolithic transmitter/receiver systems. The whole assembly is lightweight and compact. The most attractive feature of the integrated horns layer is that focal plane arrays can be fabricated easily on a single wafer. Such systems potentially offer a significantly improved spatial resolution in remote sensing, and a much greater antenna gain when implemented as arrays.

Micromachined horn antennas consist of a dipole antenna fabricated on a thin dielectric membrane (e.g. Si_3N_4) inside a pyramidal cavity etched in silicon. The micromachined array is made of a stack of silicon wafers. Arrays of additional horns can be made with a high packing density and are relatively easy to implement using a milling machine with a split block technique. The horn array can be made by using a stack of copper blocks with a gold-plated surface. An example of the receiver is shown in Figure 4.14.

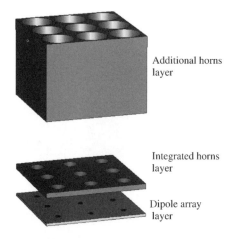

Additional horns layer

Integrated horns layer

Dipole array layer

Figure 4.14   Schematic of an array structure including a micromachined and machined horn array, the device wafer and the DC and IF connection board [5]

The integrated-circuit horn antenna has been studied using a full-wave analysis technique in Reference [6]. The circuit consists of a dipole feed evaporated on a thin dielectric membrane which is suspended in a pyramidal cavity etched in silicon or GaAs. Recently, this antenna has been applied to several millimetre and submillimetre wave applications including a double-polarised antenna design at the W-band [7]. However, the wide flare angle of the integrated-circuit horn antenna limits its useful aperture size to 1.6 λ and its gain to 13 dB. The flare angle is dictated by the anisotropic etching involved in its fabrication. For example, crystallographic constraints could limit the angle to 70° in silicon. To solve this problem, the quasi-integrated horn antenna was introduced [8], this consists of a machined small flare angle pyramidal section

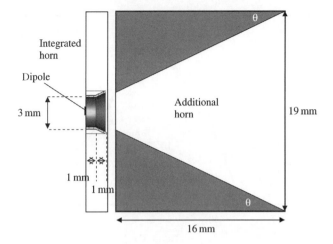

Figure 4.15   General configuration of the quasi-integrated multimode horn antenna [12]

attached to the integrated portion (Figure 4.15). The resulting structure is a simple multimode pyramidal horn with circularly symmetric patterns and low cross polarisation. This design is particularly suitable for submillimetre quasi-optical receiver applications. The minimum machined dimension involved in its geometry is around 1.5 λ, which enables its fabrication to frequencies of up to 2 THz.

A wide range of practical quasi-integrated horn antenna designs along with their radiation characteristics will now be discussed. Since a very desirable property of antennas intended for use in quasi-optical systems is the high Gaussian content of their radiated fields [9], the design methodology utilises the optimisation of the quasi-integrated horn in order to achieve maximum Gaussian coupling efficiency. The Gaussian coupling efficiency is particularly important in quasi-optical receiver applications because it directly influences the total system performance, with a significant effect on the receiver noise temperature [10]. The "Gaussian-beam" approach described here is aimed at the design of multimode horns with symmetric patterns and utilises the aperture fields directly to determine the excitation level of each mode in a simple step, instead of traditional methods using complex processing of the far-field radiation pattern [11]. Also, the large difference between the flare angles of the integrated and the machined parts of the quasi-integrated horn antenna enables the treatment of these two portions independently, resulting in a simple and efficient design approach.

The geometrical parameters for the 20 dB realisation were calculated to be ($L = 16$ mm = 3.2λ, taper angle $\theta = 24°$). Similar design parameters have been reported in Reference [13].

Figure 4.16 shows the simulation result for the E-plane and H-plane and the 10 dB power beamwidth. The indicated 10 dB beamwidth fluctuation corresponds to the variation of the beamwidth in an azimuthal far-field cut. For this purpose the aperture radius of curvature of the Gaussian beam (R_G) was obtained from the expression:

$$R_G = \frac{R_E + R_H}{2}$$

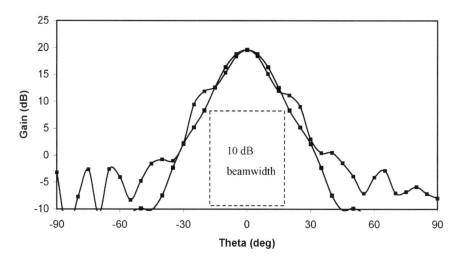

Figure 4.16 The E- (–□–) and H-plane (–■–) patterns of the 20 dB quasi-integrated horn

where R_E and R_H are the radii of curvature of the aperture field in the E-plane and H-plane cuts respectively, and were obtained from a least-squares fit to the phase of the aperture field. Also, the Gaussian-beam roll-off was calculated at the edges of the $\pm 5\%$ bandwidth using the Gaussian-beam parameters, which were calculated at the design frequency f_0.

4.4 Conical Horns and Circular Polarisation

Figure 4.17 shows an array antenna consisting of 2×2 patch array antennas with multiple microstrip feeding lines, four circular horns and four phase shifters [14]. Four square patches with truncated corners are included to generate right-hand circular polarisation (this is a simple method of generating circular polarisation which avoids two orthogonal feeder lines). The relative orientation of each patch is 90° rotated clockwise around the centre point of the array [15]. Thus, the feed to each patch has a 90° angular orientation, as shown in Figure 4.17. With this spacing, the fields generated from the four feeds are orthogonal to each other. Additionally, the four feeds are required to be fed 90° out-of-phase to achieve TM_{11} mode circular polarisation. The phase of each patch is therefore 0°, 90°, 180° and 270°, respectively, in a clockwise direction [15]. For left-hand circular polarisation (LHCP), the electric field vector will rotate in the opposite direction. The phase settings for both RHCP and LHCP are illustrated in Table 4.2.

The phase difference of 90° can be produced by either adjusting the length of the microstrip feeding lines or using a branch-line coupler. As a result, the radiation summation of the 2×2 array is right-hand circular polarisation. The prototypes of feeding lines and patch antennas are printed using a gold layer on a ceramic substrate ($\varepsilon_r = 10$). The size of a 61 GHz patch is 0.72 mm by 0.72 mm approximately.

The integrated conical horn array is placed in the ground plane to improve the gain and bandwidth of the patches, and the centre of the patch is aligned with the axis of the horn. A patch and its surrounding horn are considered together to be one radiation element. The

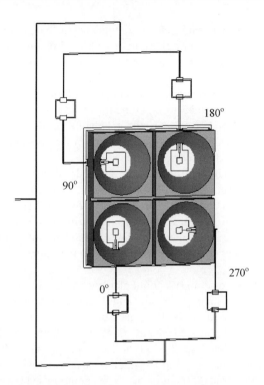

Figure 4.17 Top view of a 2 × 2 beam-steering array antenna and the size of a horn (DC power lines are not shown in the figure) [14]

Table 4.2 Summary of phase settings

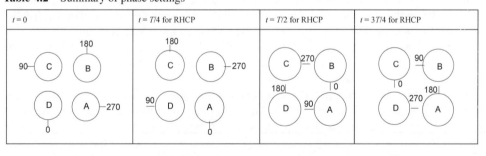

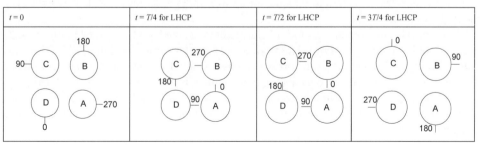

feeding lines can then connect to each patch via a tunnel at the base of the horns. All the horns are integrated into one piece made of aluminium with gold plating. Its size is 15 mm × 15 mm × 3 mm. Part of the ceramic substrate is removed so that the conical horns can have direct electrical contact with the ground plane. The integrated horn is soldered on to the ground plane for the purpose of minimizing the parasitic effect at the millimetre wave range. The radius of a horn is 1.8 mm at the bottom and 3.5 mm at the top. To achieve a small sidelobe level, the distance between adjacent elements is approximately half a wavelength. This condition limits the size and the gain of a horn. When the electrical size of the horns increases, the directivity will increase and so does the distance between the horns.

If the directivity of a single horn is too high, the radiated pattern from each element becomes independent and the array cannot form a composite beam. Therefore, the size of horn arrays has to be tailored to have an appropriate beam pattern for the design application. Each 90° phase shifter is made of two switches and two microstrip lines with different lengths. The radiation of the phase shifters and microstrip lines is shielded by a metallic cover to avoid mutual coupling. Its 61 GHz radiation pattern is shown in the Cartesian plot for different theta in Figure 4.18.

When all phase shifters are set to 0° and all elements have the same phase, the maximum directivity of RHCP signals is along the boresight direction; where the minimum directivity of left-hand circular polarisation (LHCP) signals exists (a null appears on the boresight). Figure 4.18 (a) illustrates the simulated radiation pattern with directivity of 16 dBi. The half-power beamwidth for RHCP is 20° and the first null spacing is 40°. The first sidelobe of RHCP appears at +30° and −30° and the sidelobe level is above −10 dB. The maximum gain of LHCP (cross polar) is at +50° and −50°. The compromise between the sidelobe level and the beamwidth can be varied by adjusting the distance between the elements. When the two phase shifters are at 0° and the other two are at 90°, the maximum directivity of RHCP signals tilts −10° in the theta direction where the minimum directivity of LHCP signals exists. Simulated results are shown in Figure 4.18 (b). Also, the maximum directivity of LHCP signals tilts 10° in the theta direction. Thus, it is clear that the beam direction can be steered 10° by controlling the orthogonal phase shifters. As each horn provides a directional beam with 32° half-power beamwidth, the tilt of the beam of the 2 × 2 array is distorted when the beam direction goes to 10°. To mitigate the distortion, the antenna beamwidth can be increased and the distance between the horns can be adjusted.

The simulated axial ratio (AR) as a function of elevation angle at 59, 61 and 63 GHz, respectively, is shown in Figure 4.19. The AR is less than 3 dB between −4° and 40° except at +20° and −20°, where the first nulls appear in the radiation pattern. As the antenna is designed to have 20° half-power beamwidth, circular polarisation is good within the main lobe. This makes antennas less sensitive to manufacturing tolerances and reinforces their feasibility at the millimetre wave scale.

The composite array comprises four identical horns, marked as 1, 2, 3 and 4 in Figure 4.20. A cross slot between each horn was introduced, as shown earlier in Figure 4.17. The array with a slot is marked (b) while the array (a), has no slot. The slot that is 0.4 mm wide and 1.25 mm high is designed to reduce the mutual coupling between horns. The simulated coupling coefficients between horns are plotted in Figure 4.20. This result shows that the cross slot reduces the mutual coupling by up to 10 dB. Figure 4.21 shows that the best match is found to be about −17 dB at 61.5 GHz.

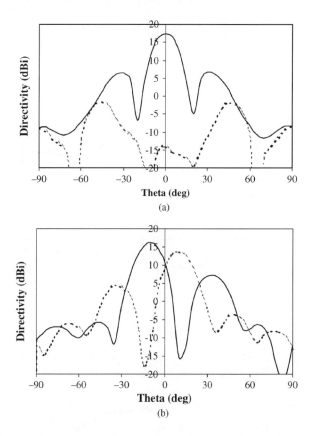

Figure 4.18 Simulated 61 GHz far-field radiation pattern of a 2 × 2 antenna array at Φ = 0° for RHCP (solid line) and LHCP (dotted line). (a) The main beam at the direction of theta = 0°. (b) The main beam at the direction of theta = −10°. (Reproduced by permission of © 2006 IEEE [14])

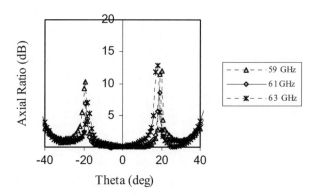

Figure 4.19 Calculated axial ratio at 59, 61 and 63 GHz as a function of elevation angle when the main beam is at the direction of theta = 0°. (Reproduced by permission of © 2006 IEEE [14])

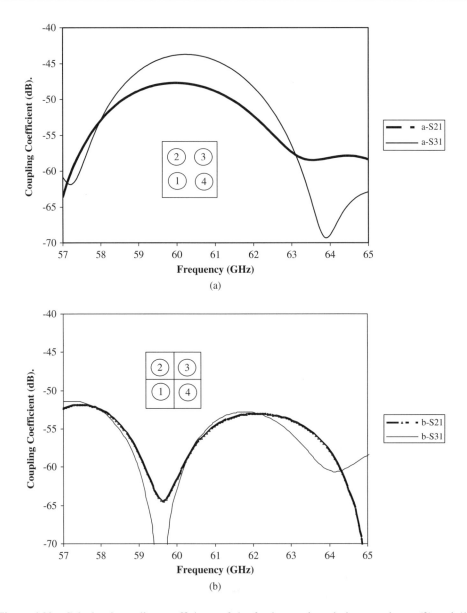

Figure 4.20 Calculated coupling coefficients of the fundamental mode between horns (2) and (1), S_{21}, and horns (3) and (1), S_{31}, in the 2×2 array. (a) Array without a slot. (b) Array with a cross slot [14]

The far-field radiation pattern was measured by a comparison method between a standard gain horn and the antenna under test. Figure 4.22 shows that the radiation pattern varies with the height of the horn. The results have been normalised with respect to the 3 mm height horn. When the horn is removed, the gain of the main beam drops by 8 dB. When the height of the

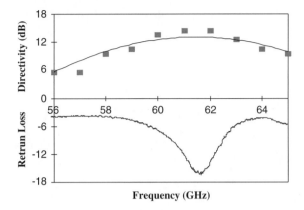

Figure 4.21 Measured directivity (square dots), simulated directivity (upper line) and measured return loss (lower line) of a 2×2 array antenna from 56 to 65 GHz. (Reproduced by permission of © 2006 IEEE [14])

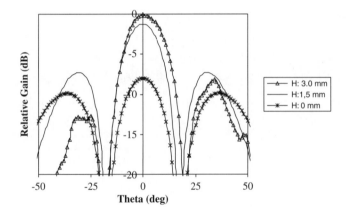

Figure 4.22 Measured 61 GHz radiation pattern at $\Phi = 0°$ with different heights (H) of horns: $H = 3.0$ mm, $H = 1.5$ mm and $H = 0$ mm (no horn). These curves are normalised to the one with $H = 3$ mm. (Reproduced by permission of © 2006 IEEE [14])

horn is 1.5 mm, the main beam has the same gain as the horn with the 3 mm height, but it should be noted that the sidelobe level is higher.

When phase shifters are all set to 0°, the radiation pattern is plotted using the Cartesian format in Figure 4.23 (a). The beam is along the boresight direction and the measured half-power beamwidth is 20°, as was predicted.

Each phase shifter consists of two 61 GHz switches and two microstrip lines in different lengths for 1-bit phase shifting. By switching between different lengths of microstrip lines, the corresponding phase is changed and therefore the beam direction is steered. As the transmission loss for the gold microstrip line on ceramics is 0.57 dB/cm at 61 GHz, the total length of the antenna feeding line is 7 cm and therefore has approximately 4 dB losses. In addition, each

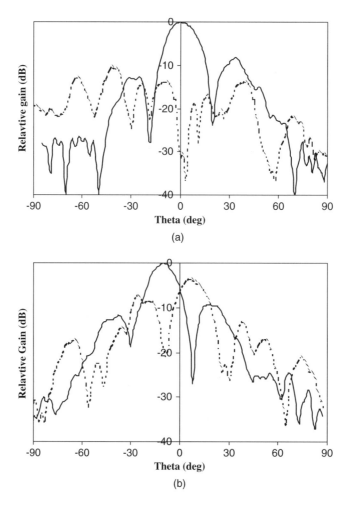

Figure 4.23 A 61 GHz far-field radiation pattern measured result at $\Phi = 0°$ for RHCP (solid line) and LHCP (dotted line). (a) Main beam at the direction of theta $= 0°$. (b) Main beam at the direction of theta $= -10°$.(Reproduced by permission of © 2006 IEEE [14])

element has two switches for making the phase shift. The 61 GHz switches from Northrop Grumman Space Technology and its wire bonding, are measured to have 3 dB losses for each phase shifter on the ceramic substrate. The antenna gain is measured to be up to 7 dBi at 61 GHz. After the de-embedment of feeding line connector discontinuity and the short-open-through-line calibration between 56 and 65 GHz, the directivity of 14 dBi is measured at 61 GHz as shown in Figure 4.21.

As the antenna prototype is implemented with feeding networks and MMICs, on top of the ceramic substrate, radiation from feeding networks and MMICs can destroy the cross polarisation performance. It can be seen from the measured radiation pattern in Figure 4.23 (a), that cross polarisation on the z axis is reduced to approximately 15 dB.

When phase shifters are set to 0°, 0°, 90° and 90°, the beam direction can have a 10° offset as shown in Figure 4.23 (b). The maximum RHCP signal occurs at the direction of −10°, while the LHCP signal has a maximum at +10°.

During the measurement, surface waves can be diffracted by, or coupled to feeding lines or phase shifters on the ceramics substrate. The diffraction of surface waves can cause undesirable effects on the sidelobe level, polarisation or main beam shape. Therefore, it is important to shield the feeding networks and MMICs to minimize the effect of surface waves.

The antenna can be further optimised when a multilayer structure is adopted [16]. Feeding lines and MMICs can be designed as separate layers from patches in order to reduce the coupling effect. The lengths of the feeding network are shortened and the transmission loss will therefore be reduced.

Such multihorns can also be applied to the already well-known strip slot foam inverted patch antenna (SSFIP) [17, 18], which is easy to manufacture using classical technologies and exhibit good electromagnetic characteristics that are easy to match, as well as large bandwidth and good gain. The proposed structure can be used for high radiation efficiency antennas in the millimetre wave band since the undesirable surface waves are inherently suppressed with the use of a metallic horn cavity configuration. Without horns, the surface wave can drastically alter the behaviour of an antenna by reducing its bandwidth, deteriorating its radiation pattern, or lowering its efficiency.

Possible solutions to reduce the surface-wave excitation are the use of low-index dielectrics, which excite mainly space waves, as circular patches with appropriate radii or as membranes supporting rectangular patches over an air cavity or a low-index dielectric material.

The general layout of the antenna and the materials used are presented in Figure 4.24.

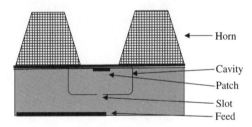

Figure 4.24 Design of the integrated horn antenna. The antenna is composed of two wafers: a silicon wafer with a horn, a membrane supporting a patch and an auxiliary hole for the electric contact with the second wafer and a Pyrex wafer with a cavity, a slot and a feeding line

A reasonable way to analyse the antenna of Figure 4.24 is to consider it as being a succession of connected metallisation embedded in a multilayered medium. Thus, the problem falls into the category of microstrip antennas with vertical connections, embedded in stratified media. This subject has been in the scope of research for many years now, so is already covered in the literature for either horizontal sources [19–21] or, more recently, for vertical sources [22–24].

As shown in Figure 4.24, the antenna is composed of two parts having complementary roles: the first one, a Pyrex wafer associated with an annular line resonance, supports the excitation of the antenna, whereas the second one, a silicon wafer, represents the radiating part. Previous analysis of SSFIP antennas [25] has shown that special attention should be paid to the parameters (line–slot–patch) since they are of primary importance for the adaptation of the device. The large number of simulations run has shown that the size of the cavity, once adjusted, has second-order influence on the frequency of operation.

4.5 Tilt Horn

When a horn is tilted off-axis, the direction of the radiation main beam is also tilted. This method can be used to control the beam direction, which is especially useful for beam-switching antennas. Figure 4.25 shows a patch-fed horn 20° off-axis in the x direction; its radiation pattern at 60 GHz can be found in Figure 4.26.

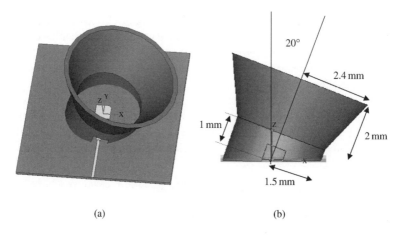

(a) (b)

Figure 4.25 Patch-fed tilt horn with 20° off-axis in the x direction

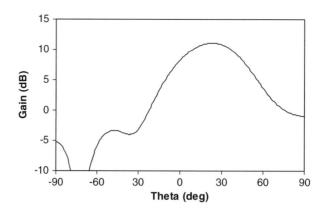

Figure 4.26 Radiation pattern for the tilt horn

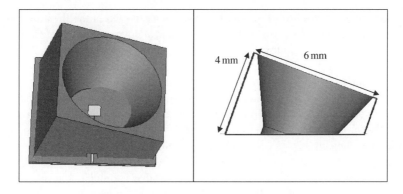

Figure 4.27 Tilt horn inside a metal block

Figure 4.28 Cut view of a multihorn antenna

For manufacturing convenience, the horn can be embedded in a metal block, as shown in Figure 4.27. Therefore, it is possible to have multihorn manufacturing in one metal block, as shown in Figure 4.28. Three horns are designed in one metal board, each of them having different radiation coverage as a conformal antenna. By applying the beam-switching technique, this antenna can provide high gain and beam switching with low manufacturing cost.

For a multilayer material such as low-temperature co-fired ceramics (LTCC), a quasi-horn antenna can be constructed by stacking multiple layers with different sizes of windows, as shown in Figure 4.29.

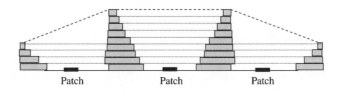

Patch   Patch   Patch

Figure 4.29 Multihorn in a multilayer LTCC configuration

4.6 Dielectric Sectoral Horn

Dielectric antennas are of great importance because of their low loss, high gain, light weight, their feasibility of obtaining shaped beams, ease of fabrication, etc. Solid and hollow dielectric horn antennas have received special attention due to their increased directivity and high gain compared to metallic horns [26]. Few studies have been reported in the literature describing results from rectangular hollow dielectric horn antennas. However, for the record, included here is the description of a hollow dielectric horn antenna capable of producing a flat-top radiation pattern with low sidelobe levels and cross polarisation in the H-plane [27].

A diagram of the strip-loaded hollow dielectric H-plane horn antenna is shown in Figure 4.30 [27]. The dielectric part of the horn is fabricated using low-loss dielectric polystyrene, and was fixed at the end of an open metallic waveguide. A carefully tapered dielectric rod launcher is placed at the throat of the antenna in order to reduce the feed-end discontinuity. The taper length inside the waveguide was optimised for a minimum VSWR. Two thin metal strips of length *l* were placed on the H-walls of the horn, which significantly modifies the aperture field of the horn, changing the radiation pattern considerably. The sidelobe levels and the half-power beamwidth (HPBW) of the E- and H-plane patterns can be adjusted by changing the strip length. More details about the dielectric antenna can be found in the next chapter.

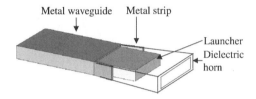

Figure 4.30   Schematic diagram of a strip-loaded hollow dielectric H-plane sectoral horn antenna [27]

References

[1] R.E. Collin, *'Foundations for Microwave Engineering'*, IEEE Press, 1992.

[2] Yin Xinghui, 'A New Design Technique for Millimetre Wave Horn Antennas', ICMMT Microwave and Millimetre Wave Technology Proceedings of International Conference, 1998, pp. 380–381.

[3] D. J. Edwards *et al.*, 'Electronic Tracking System for Microwave Antenna', US Patent 4704611, 1987.

[4] G. M. Rebeiz, D. P. Kasilingam, Y. Guo, P. A. Stimson and D. B. Rutledge, 'Monolithic Millimetre-Wave Two-Dimensional Horn Imaging Arrays'. *IEEE Transactions on Antennas and Propagation*, **38**(9), September 1990, 1473–1482.

[5] Gert de Lange, Konstantinos Konistis and Qing Hua, 'A 333 Millimetre-Wave Micromachined Imaging Array with Superconductor–Insulator–Superconductor Mixers', *Applied Physics Letters*, **75**(6), 9 August 1999, 868–870.

[6] G. V. Eleftheriades, W. Y. Afi-Ahmad, L. P. B. Katehi and G. M. Rebeiz, 'Millimetre-Wave Integrated-Horn Antennas. Part I – Theory and Part II – Experiment', *IEEE Transactions on Antennas and Propagation*, **39**, November 1991, 1575–1586.

[7] W. Y. Ali-Ahmad, G. M. Rebeiz, H. Dave and G. Chin, '802 GHz Integrated Horn Antennas Imaging Array', *International Journal of Infrared Millimetre Waves*, **12**(5), 1991.

[8] G. V. Eleftheriades, W. Y. Afi-Ahmad and G. M. Rebeiz, 'A 20-dB Quasi-Integrated Horn Antenna', *IEEE Microwave Guided Wave Letters*, **2**, February 1992, 73–75.

[9] P. F. Goldsmith, 'Quasi-Optical Techniques at Millimeter and Submillimeter Wavelengths', in *'Infrared and Millimetre Waves'*, Vol. 6, Academic Press, New York, 1982, pp. 243–277.

[10] R. Padman, J. A. Murphy and R. E. Hills, 'Gaussian Mode Analysis of Cassegrain Antenna Efficiency', *IEEE Transactions on Antennas and Propagation*, **35**, October 1987, 1093–1103.

[11] H. M. Pickett, J. C. Hardy and J. Farhoomand, 'Characterization of a Dual-Mode Horn for Submillimeter Wavelengths', *IEEE Transactions on Microwave Theory and Techniques*, **32**, August 1984, 933–937.

[12] George V. Eleftheriades and Gabriel M. Rebeiz, 'Design and Analysis of Quasi-Integrated Horn Antennas for Millimeter and Submillimeter-Wave Applications', IEEE Transactions on Microwave Theory and Techniques, **41**(6/7), June/July 1993, 954–965.

[13] P. D. Potter, 'A New Horn Antenna with Suppressed Sidelobes and Equal Beamwidths', *Microwave Journal*, **VI**, June 1963, 71–78.

[14] K. Huang and Z. Wang, 'Millimetre-Wave Circular Polarized Beam-Steering Antenna Array for Gigabit Wireless Communications', *IEEE Transactions on Antennas and Propagation*, **54**(2), February 2006, 743–746.

[15] K. Huang, S. Koch and M. Uno, 'Circular Polarized Array Antenna', European Patent EP1564843.

[16] R. Kulke, C. Günner, S. Holzwarth, J. Kassner, A. Lauer, M. Rittweger, P. Uhlig and P. Weigand, '24 GHz Radar Sensor Integrates Patch Antenna and Frontend Module in Single Multilayer LTCC Substrate', 15th European Microelectronics and Packaging Conference (IMAPS), Brugge, 12–15 June 2005, pp. 239–242.

[17] J.-F. Zucher and F .E. Gardiol, 'SSFIP: A Global Concept for High Performance Broadband Planar Antennas, *Electronics Letters*, **24**, 1988, 1433–1435.

[18] J.-F. Zurcher and F. E. Gardiol, *'Broadband Patch Antennas'*, Artech House, Norwood, Massachusetts, 1995.

[19] J. R. Mosig and F. Gardiol, 'General Integral Equation Formulation for Microstrip Antennas and Scatterers', *Proceedings of the Institution of Electrical Engineers*, **132**, 1985, 424–432.

[20] J. R. Mosig, 'Arbitrarily Shaped Microstrip Structures and Their Analysis with a Mixed Potential Integral Equation', *IEEE Transactions on Microwave Theory and Techniques*, **36**,1988, 314–323.

[21] Y. L. Chow, N. Hojjat, S. Safavi-Naeini and R. Faraji-Dana, 'Spectral Green's Functions for Multilayer Media in a Convenient Computational Form', *Proceedings of the Institution of Electrical Engineers*, **145**, 1998, 85–91.

[22] N. Kinayman and M. I. Aksun, 'Efficient Use of Closed-Form Green's Functions for the Analysis of Planar Geometries with Vertical Connections', *IEEE Transactions on Microwave Theory and Techniques*, **45**, 1997, 593–603.

[23] T. M. Grzegorczyk and J. R. Mosig, 'Full-Wave Analysis of Antennas Containing Horizontal and Vertical Metalizations Embedded in Multilayered Media', *IEEE Transactions on Antennas and Propagation*, **51**, 2003, 3047–3054.

[24] R. F. Harrington, *'Field Computation by Moment Method'*, IEEE Press, New York, 1993.

[25] Y. Brand, 'Antennes Imprimees SSFIP: de l'Element Isole au Reseau Planaire', PhD Dissertation, Ecole Polytechnique Federale 'de Lausanne, Switzerland, 1996.

[26] V. P. Joseph, S. Mathew, J. Jacob, U. Ravindranath and K. T. Mathew, 'Radiation Characteristics of Strip Loaded Hollow Dielectric Ž. E-Plane Sectoral Horn Antennas', *Electronics Letters*, **33**, 1997, 2002–2004.

[27] V. P. Joseph, S. Biju Kumar and K. T. Mathew, 'Strip-Loaded Hollow Dielectric H-Plane Sectoral Horn Antennas for Square Radiation Pattern', *Microwave Optical Technology Letters*, **29**, 2001, 45–46.

5

Dielectric Antennas

Dielectric rod antennas have been used extensively as an endfire radiator for many years, and a considerable number of theoretical and experimental studies for this type of antenna have been published at microwave and millimetre wave frequencies [1]. This type of antenna has the attractive features of light weight and low cost. Therefore presented here are the basic design considerations, properties and feeding techniques. The chapter is divided into seven sections. Section 5.1 introduces the dielectric resonator antenna. Section 5.2 describes the dielectric rod antenna. Section 5.3 continues with maximum gain antennas, and then dual rods are discussed in Section 5.4. Section 5.5 presents the basic features of a microstrip patch-fed dielectric antenna, while Section 5.6 discusses a compensation method for a dielectric array with different phase delays to the elements. Finally, Section 5.7 describes techniques for the optimisation of the performance of a rod antenna.

5.1 Dielectric Resonator Antennas

Dielectric resonators appeared in the 1970s in work that led to the miniaturization of active and passive microwave components, such as oscillators and filters [2]. In a shielded environment, the resonators built with dielectric resonators can reach an unloaded Q factor of 20 000 at frequencies of 20 GHz.

The principle of operation of the dielectric resonator can best be understood by studying the propagation of electromagnetic waves on a dielectric rod waveguide (Reference [2], Chapter 3). The mathematical description [3] and the experimental verification [4] of the existence of these waves have been known for a considerable time. Their wide application was prompted by the introduction of optical fibres in communications systems.

One of the attractive features of a dielectric resonator antenna (DRA) is that it can assume any one of a number of simple shapes, the most common being ones with circular or rectangular cross-sections, as shown in Figure 5.1. Over the years the frequency range of interest for many systems has gradually progressed upwards to the millimetre and near-millimetre range (100–300 GHz). At these frequencies, the conductor loss of metallic antennas becomes severe and the efficiency of such antennas is significantly reduced. Conversely, the only loss for a dielectric resonator antenna is that due to the imperfect dielectric material (the loss tangent

Millimetre Wave Antennas for Gigabit Wireless Communications   Kao-Cheng Huang and David J. Edwards

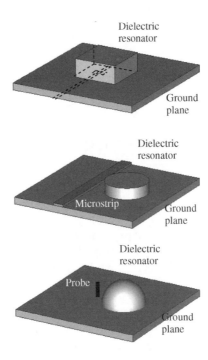

Figure 5.1 Typical dielectric resonator antennas and feeding mechanisms: (a) aperture feed, (b) microstrip feed and (c) probe feed

of the material), which can be very small in practice. Some example materials are shown in Table 5.1. After the cylindrical dielectric resonator antenna had been studied [5], the rectangular and hemispherical dielectric resonator antennas were subsequently investigated [6]. This work created the foundation for future investigations of the dielectric resonator antennas. Other shapes were also studied, including cylindrical-ring [7] and spherical-cap dielectric resonator antennas. It was found that dielectric resonator antennas operating at their fundamental modes radiate like a magnetic dipole, independent of their shapes. A few dielectric resonator suppliers are listed in Table 5.1; the materials and dielectric constants of the dielectric resonators are also shown.

As with all bounded systems, the field distribution within the structure is defined by the mode. The mode of propagation is a solution to the wave equation that satisfies both the boundary conditions and the excitation (feeding) method. Some of the lowest modes of propagation on dielectric rod waveguides are shown in Figures 5.2 to 5.4.

The first index denotes the number of full-period field variations in the tangential direction and the second one the number of radial variations. When the first index is equal to zero, the electromagnetic field is circularly symmetric. In the cross-sectional view, the field lines can be either concentric circles (like, for example, the E-field of the TE_{01} mode) or the radial straight line (like, for example, the H-field of the same mode). For higher modes, the pure transverse electric or transverse magnetic fields cannot exist, so both electric and magnetic fields must have non-vanishing tangential components. Such modes are called hybrid electromagnetic (HEM); the lowest of them being HEM_{11}.

Table 5.1 Some dielectric resonator suppliers, along with the materials and dielectric constants of their dielectric resonators [8, 9]

	Type	Dielectric constant	Dielectric loss tangent (e')	Note
Countis	CD-6	6.3	< 0.00015	$MgO–SiO_2$
Laboratories	CD-9	9.5	< 0.00015	$MgO–SiO_2–TiO_2$
	CD-13	13.0	< 0.00015	$MgO–TiO_2–SiO_2$
	CD-15	15.0	< 0.00015	$MgO–TiO_2$
	CD-16	16.0	< 0.00015	$MgO–TiO_2$
	CD-18	18.0	< 0.00015	$MgO–CaO–TiO_2$
	CD-20	20.0	< 0.00015	$MgO–CaO–TiO_2$
	CD-30	30.0	< 0.0002	$MgO–CaO–TiO_2$
	CD-50	50.0	< 0.0005	$MgO–CaO–TiO_2$
	CD-100	100.0	< 0.0008	$MgO–CaO–TiO_2$
	CD-140	140.0	< 0.0010	$MgO–CaO–TiO_2$
Emerson and	ECCOSTOCK® HT0003	2.2	< 0.0003	
Cuming	ECCOSTOCK® CPE	2.4	0.0001	Low-loss/low-dielectric thermoset materials
Microwave Products	ECCOSTOCK® 0005	2.54	0.0005	
	ECCOSTOCK® HiK	3 to 15	< 0.002	
	ECCOSTOCK® HiK500F	3 to 30	< 0.002	Low-loss materials
	ECCOSTOCK® CK	1.7 to 15	< 0.002	Low-loss moulded products

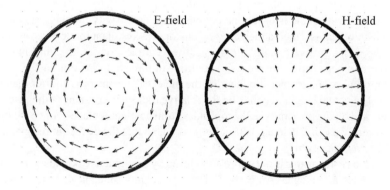

Figure 5.2 Mode TE_{01} on a dielectric rod waveguide. Left: E-field, right: H-field

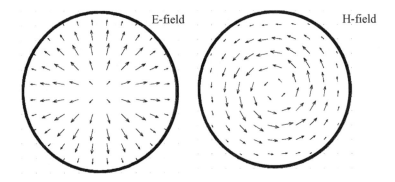

Figure 5.3   Mode TM_{01} on a dielectric rod waveguide. Left: E-field, right: H-field

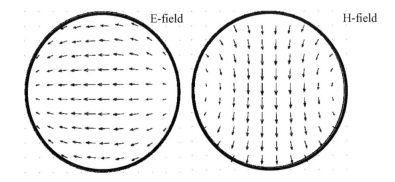

Figure 5.4   Mode HEM_{11} on a dielectric rod waveguide. Left: E-field, right: H-field

The fields of a cylindrical rod can be expressed in terms of Bessel functions, which determine the wavelength and the propagation velocity of these waves. When only a truncated section of the dielectric rod waveguide is used, one obtains a resonant cavity in which standing waves appear. Such an object is called a dielectric resonator. The resonant mode $TE_{01\delta}$ is most often used in shielded microwave circuits. In classical waveguide cavities, the third index is used to denote the number of half-wavelength variations in the axial direction of the waveguide. Here, the third index, δ , denotes the fact that the dielectric resonator is shorter than one-half wavelength. The actual height depends on the relative dielectric constant of the resonator and the substrate, and on the proximity to the top and bottom conductor planes. Since the numerical value of δ is rarely used, this index is usually ignored, so that the dielectric resonator is often specified by two indices only.

When a dielectric resonator is not entirely enclosed by a conductive boundary, it can radiate, and so it becomes an antenna. An early dielectric resonator antenna was successfully built and is described in Reference [10], while the rigorous numerical solution was published in Reference [11]. Review treatments of dielectric resonator antennas can be found in References [12] and [13].

As shown in Figure 5.5, the dielectric resonator element is placed on a ground plane, and a short electric probe penetrates the plane into the resonator. The probe is located off centre, close to the perimeter of the resonator. The radiation occurs mainly in the broadside direction (i.e. radially) and is linearly polarised.

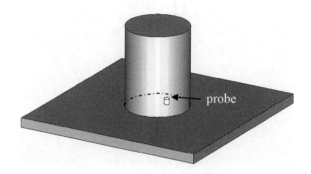

Figure 5.5   Dielectric resonator antenna fed with a coaxial probe

The numerical analysis of the dielectric resonator antenna started as an attempt to determine the natural frequencies of various modes in an isolated dielectric resonator, without any other scattering object in its vicinity and without any excitation mechanism. It was found that the resonant frequencies are complex-valued:

$$f_{m,n} = \sigma_{m,n} + j\omega_{m,n} \tag{5.1}$$

where σ represents the in-phase (real) component (which is the lossy component) and ω is the out-of-phase component (imaginary) with respect to the excitation.

Each particular solution corresponds to a resonant m, n type mode that satisfies all the boundary and continuity conditions. For rotationally symmetric resonators, subscript m denotes the number of azimuthal variations and subscript n denotes the order of appearance of modes in the growing frequency direction.

The fact that the resonant frequency has a non-vanishing real part signifies that such a mode would oscillate in an exponentially decaying manner if it was initially excited by an abrupt external stimulus. The ratio of the real to the imaginary part of the resonant frequency is the radiation Q factor of the mode:

$$Q_r = -\frac{\omega_{m,n}}{2\sigma_{m,n}} \tag{5.2}$$

The negative sign arises from the observation that all passive circuits have their natural frequencies located on the left-half complex plane, so $\sigma_{m,n}$ is itself a negative number. The natural frequencies and the radiation Q factors of the modes are given as follows [14]:

TE$_{01\delta}$ mode:

$$f_0 = \frac{2.921 c \varepsilon_r^{-0.465}}{2\pi A} \left[0.691 + 0.319 \left(\frac{A}{2H} \right) - 0.035 \left(\frac{A}{2H} \right)^2 \right]$$

$$Q = 0.012 \varepsilon_r^{1.2076} \left[5.270 \left(\frac{A}{2H} \right) + 1106.188 \left(\frac{A}{2H} \right)^{0.625} e^{-1.0272(A/2H)} \right]$$

HE$_{11\delta}$ mode:

$$f_0 = \frac{2.735 c \varepsilon_r^{-0.436}}{2\pi A} \left[0.543 + 0.589 \left(\frac{A}{2H} \right) - 0.050 \left(\frac{A}{2H} \right)^2 \right]$$

$$Q = 0.013 \varepsilon_r^{1.202} \left[2.135 \left(\frac{A}{2H} \right) + 228 \left(\frac{A}{2H} \right) e^{-2(A/2H)+0.11(A/2H)^2} \right]$$

TM$_{01\delta}$ mode:

$$f_0 = \frac{2.933 c \varepsilon_r^{-0.468}}{2\pi A} \left\{ 1 - \left[0.075 - 0.05 \left(\frac{A}{2H} \right) \right] \left[\frac{\varepsilon_r - 10}{28} \right] \right\}$$

$$\left\{ 1.048 + 0.377 \left(\frac{A}{2H} \right) - 0.071 \left(\frac{A}{2H} \right)^2 \right\}$$

$$Q = 0.009 \varepsilon_r^{0.888} e^{0.04} \varepsilon_r \left\{ 1 - \left[0.3 - 0.2 \left(\frac{A}{2H} \right) \right] \left[\frac{38 - \varepsilon_r}{28} \right] \right\}$$

$$\left\{ 9.498 \left(\frac{A}{2H} \right) + 2058.33 \left(\frac{A}{2H} \right)^{4.322} e^{-3.5(A/2H)} \right\}$$

For given dimensions and a given dielectric constant, the numerical solution can determine the resonant frequency and the radiation Q factor. Such computed data can be fitted to convenient analytic expressions [15]. For instance, the resonant frequency of the HEM$_{11}$ mode of an isolated dielectric resonator radiator, of radius a and height h, can be approximated by the following expression:

$$k_0 a = (1.6 + 0.513x + 1.392x^2 - 0.575x^3 + 0.088x^4)/\varepsilon_r^{0.42} \tag{5.3}$$

Here, k_0 is the free space propagation constant and $x = a/h$. Similarly, the values of Q_r for the same mode can be calculated from:

$$Q_r = x \varepsilon_r^{1.2} (0.01893 + 2.925 e^{-2.08x(1-0.08x)}) \tag{5.4}$$

An alternative way of exciting the HEM$_{11}$ mode in the dielectric resonator antenna is by the microstrip-slot mechanism shown in Figure 5.6. Instead of a coaxial line, the feeding is done by a microstripline that runs below the ground plane. There is a narrow slot (aperture) in the ground plane (which is the upper layer here) for coupling the microstrip to the dielectric resonator antenna.

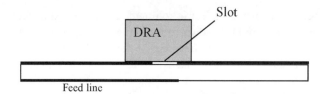

Figure 5.6 Dielectric resonator antenna with a microstrip-slot excitation

There is a gain enhancement technique available for the dielectric resonator antenna by using a surface mount quasi-planar horn. The geometry of the antenna is shown in Figure 5.7. In the rod horn structure, the aperture-coupled DRA now works as a feed to the surface-mounted horn antenna. Note that the rectangular DRA is located over the centre of the rectangular slot (aperture feed) in the ground and is excited by a 50 Ω microstripline feed. If the surface-mounted horn is made of thin copper/aluminium sheet and is supported on a foam structure, the gain can be further improved.

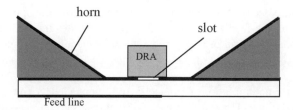

Figure 5.7 Aperture-coupled rectangular dielectric resonator antenna with horn mounting

5.2 Dielectric Rod Antennas

The difficulties inherent in constructing antennas at millimetre wave frequencies have recently spurred further interest in dielectric antennas. It has long been known that dielectric rod surface-wave antennas are good directional radiators in the endfire direction, and rod antennas of circular cross-section have been investigated analytically and experimentally [16, 17]. However, the application of these antennas has been limited due to their relatively low gain. Recent commercial interest in developing millimetre wave dielectric circuits created the additional need for low-cost antennas that can be easily integrated into an entire system [18].

Dielectric antennas of rectangular cross-section that are compatible with the dielectric waveguides of a millimetre wave integrated circuit will be discussed next. Presented in this section are experimental results for tapered rod antennas designed for low sidelobes and for maximum gain. These antennas were constructed with a low-loss material with a relative permittivity of 3 at 60 GHz.

Dielectric rod antennas have been used as endfire radiators for many years [19–28]. Experimental studies have been conducted both at microwave and millimetre wave frequencies [23–28] . Despite the extensive use of these antennas, no exact design procedure exists for them [28]. Theoretical methods usually involve simplifications and only provide general design guidelines [19–22].

The radiation behaviour of the dielectric rod antenna can be explained by the so-called discontinuity radiation concept [19], in which the antenna is regarded as an array composed of two effective sources at the feed and free ends of the rod. Part of the power excited at the feed is converted into guided-wave power, and is transformed into radiation power at the free (open) end.

The remaining power is converted into unguided-wave power radiating near the feed end. Thus, the directivity of the dielectric rod antenna is characterised by the directivities generated by these two effective sources. However, there is the problem of quantitatively computing the radiation fields generated from the discontinuities at the feed and free ends.

To date, the dielectric rod has received much attention in terms of waveguide analysis as well as the antenna analysis. Of prime importance is knowledge of the eigenmodes is in the design of a dielectric waveguide circuit. Numerous approaches have been proposed for this important issue [29–34]. Note that the methods developed in References [32] to [34] are based on Yee's mesh. The use of Yee's mesh has the advantage that the obtained eigenmode fields can directly be used in the finite difference time-domain (FDTD) method [35, 36].

The field near the rod can be decomposed into guided and unguided waves. Using the obtained solutions, a feed pattern can be calculated that corresponds to the directivity generated by the effective source at the feed end and a terminal pattern corresponding to that at the free end. Superposing the feed and terminal patterns, generates the radiation pattern of the dielectric rod antenna. The gain of a long rod antenna is calculated by superposing the feed and terminal patterns. The details of numerical analysis for rod antennas can be found in Reference [37].

Figure 5.8 shows the configuration of a dielectric rod antenna fed by a rectangular waveguide with a planar ground plane. The rod is made of Teflon (registered trademark of PTFE). It is assumed that the metallic waveguide and the ground plane are perfectly conducting and that the dielectric rod is a lossless medium [38–40]. The basic parameters of such a rod antenna are shown in Figure 5.9.

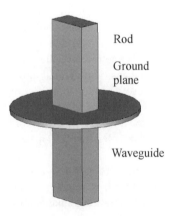

Figure 5.8 Overall geometry of a dielectric rod antenna

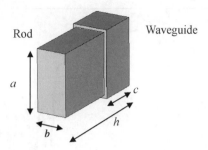

Figure 5.9   Rectangular rod ($a = 4.4\,$mm, $b = 2.2\,$mm, $c = 3\,$mm, $h = 8\,$mm)

The dimensions of the base area determine the resonant frequency of a rod antenna. In this instance the width b is half as long as the length a, i.e. $a = 2b$. When the dimension of b is changed, so is the resonant frequency, as shown in Figure 5.10.

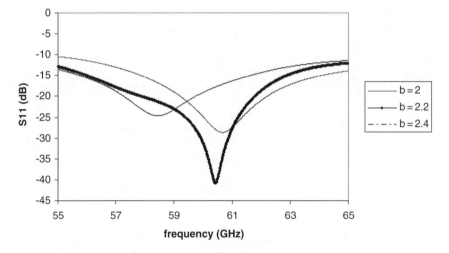

Figure 5.10   Resonant frequency of a rod antenna

The gain of the rod antenna increases as the height of the rod increases. Figure 5.11 shows the gain difference for rods of height $h = 6$, 8 and 12 mm. When the gain increases, the sidelobe may also increase. Therefore, careful tuning is needed to give a good performance.

At some point, the gain will be saturated (i.e. reaches a maximum value) and does not undergo further increases, even though the height h is increasing as shown in Figure 5.12. Therefore, it is important to know the maximum gain that can be achieved for a rod antenna (see Section 5.3). Also from this value the performance for a compact rod antenna can be assessed.

Figure 5.13 shows E-plane patterns for an 8 mm long antenna designed for 60 GHz. The gain of this antenna is 12 dB and the half-power beamwidth is 50° in the E-plane. If the rod

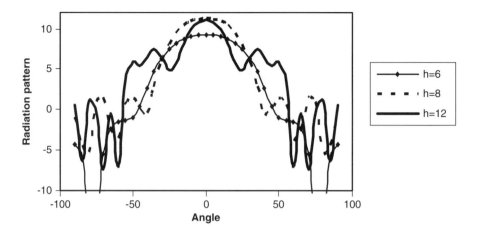

Figure 5.11   Radiation patterns for different rod heights of 6, 8 and 12 mm

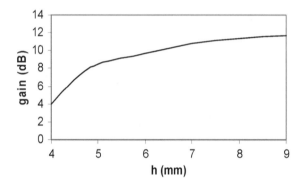

Figure 5.12   Gain curve versus height of the rod antenna in Figure 5.9 ($a = 4.4$ mm, $b = 2.2$ mm, $c = 3$ mm)

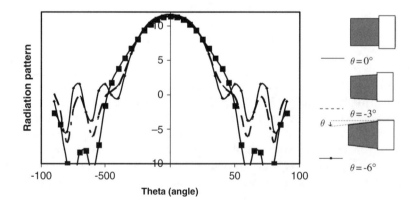

Figure 5.13   E-plane radiation pattern for E-plane tapered rods with different taper angles

does not have a tapered angle, the sidelobes, which have a broad envelope, are as high as 2 dB. When the rod is tapered, Figure 5.13 shows that the sidelobe of the rod antenna is also reduced accordingly. When the taper angle is −6°, the sidelobe at theta = 70° is reduced by approximately 10 dB.

The three designs of tapered dielectric rod antennas considered in this section are shown in Figure 5.14. All of these antennas are fed by a metal waveguide and are matched to the waveguide by a launching horn developed by Trinh *et al.* [41]. It should be noted that without the launching horn, the feed point would radiate heavily, causing many sidelobes in the far-field pattern, whose envelope is the radiation pattern of the waveguide aperture.

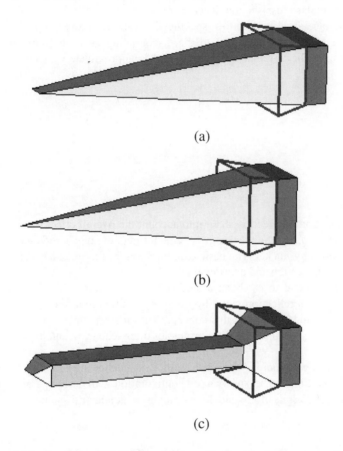

(a)

(b)

(c)

Figure 5.14   Examples of dielectric rod antennas with rectangular cross-sections [1]

A rod tapered in only the E-plane (Figure 5.14 (a)), which is easier to manufacture and mechanically stronger than an antenna tapered in both the E- and H-planes, shows no sacrifice in antenna characteristics and sometimes even shows a slight improvement. However, a rod tapered in only the H-plane is always worse than one tapered in both planes.

A second type of rod antenna is shown in Figure 5.14 (b). This rod is tapered linearly to a point in both the E- and H-planes [1]. Experimental measurements for this antenna show a sidelobe level lower than -25.5 dB from the main beam and a gain of 17.0 dB, which is slightly lower than that of the feed horn by itself (17.3 dB). Its half-power beamwidth is 30.0°, which is larger than that of the feed horn (26.9°); however, the 26 dB beamwidth of this antenna is only 72° compared with 115° for the horn. Small sidelobe characteristics and a steep sidelobe roll-off are both features of a relatively short tapered rod antenna with a feed horn.

5.3 Maximum Gain Rod Antennas

Figure 5.14 (c) shows Zucker's design principles for maximum gain antennas [17] applied to a rod of rectangular cross-section. Zucker observed that radiation from the surface-wave structure takes place at discontinuities, specifically at the feed and terminal points. He also showed that the radiation pattern due to the discontinuity at the termination is calculated by integrating over the terminal aperture S_t, and is approximately expressed as $1/\Psi$ [1], where:

$$\Psi = \frac{1}{2} k_0 H(r - \cos\theta) \tag{5.5}$$

and:

$$r = \frac{k_2}{k_0}$$

$$H = \text{height of the rod}$$

If the radiation from the feed is taken into account the overall effect is usually to sharpen the mainlobe. Based on this fact, a design scheme for maximum gain rod antennas was experimentally developed by Zucker. The basic configuration for this antenna is characterised by a feed taper, a straight section and a terminal taper.

The feed taper is said to establish a surface wave along the straight section while the terminal taper reduces reflection caused by an abrupt discontinuity (a recognised approach to impedance matching). Applying this concept, Zucker's principles are typically useful for designing a very long rod antenna; these principles often lead to an antenna with a very small cross-section.

Zucker's [17] design principles for a maximum gain antenna are adopted for rods of rectangular cross-section. Figure 5.15 shows a rod configuration with design parameters that might be useful for actual designs of low-sidelobe or high-gain dielectric antennas.

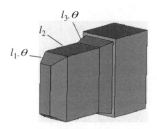

Figure 5.15   Basic rod configuration and taper angle θ, which is the same for both tapers

The simulation result of Figure 5.15 is shown in Figure 5.16. When the taper angle θ is changed, the radiation pattern changes accordingly. When the taper angle is $0°$ and $-10°$, the gain of the radiation pattern is almost the same. When the taper angle is $-20°$, the gain is reduced by 2 dB and the sidelobes are also minimised at $70°$ and $-70°$.

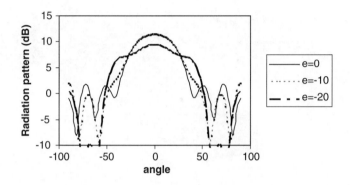

Figure 5.16 Radiation patterns for different taper angles ($l_1 = l_3 = 1$ mm, $l_2 = 3$ mm)

5.4 The Dual Rod Antenna

In gigabit wireless communication systems, two rod antennas can be used, one being responsible for the transmission and the other reception, and the performance of the whole system depends largely on obtaining minimum interference between the two antennas. As it is important to reduce the space occupied by the antennas in such systems, it is clear that the distance between the two rods should be optimised in order to achieve a minimum acceptable coupling between them. In [42], a method for calculating the inter-rod coupling coefficient by means of an optimised model for each antenna was proposed, in which the mutual and self-admittance were determined. The method also allowed the radiation patterns to be obtained.

Figure 5.17 shows the system the coupling coefficient can be defined as the ratio of the power coupled to one of the rods to the input power of the other. In order to investigate this coupling, the system can be represented by an equivalent circuit. A single antenna can be considered as a lossy load represented by a complex admittance and the coupling between the two rods is accounted for by means of the mutual admittance. In this way, the symmetric coupled antenna system can be considered as a two-port network represented by the π-type equivalent circuit of Figure 5.18. In this circuit, Y_0 stands for the self-admittance of each antenna and Y_1 and Y_2 characterise the coupling effects. The coupling coefficient is deduced from the transfer function of the two-port network i.e.:

$$C(\text{dB}) = 20 \log \left| \frac{V_{\text{out}}}{V_{\text{in}}} \right| = 20 \log \left| \frac{Y_1}{Y_0 + Y_1 + Y_2} \right| \qquad (5.6)$$

As illustrated in Figure 5.17, a symmetry plane separates the structure into two parts.

When this symmetry plane is an electric wall, the input equivalent admittance will be the odd admittance given by:

$$Y_{\text{odd}} = Y_0 + Y_2 + 2Y_1 \qquad (5.7)$$

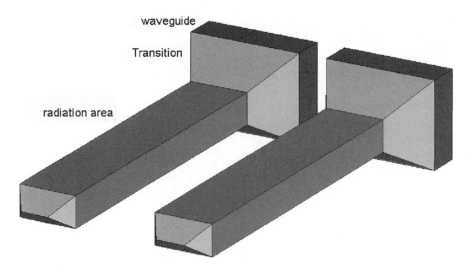

Figure 5.17 Coupled rod antenna system. (Reproduced by permission of © 1982 IEEE [42])

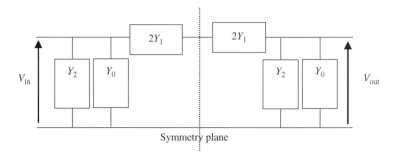

Figure 5.18 The π-type equivalent circuit Y_0, self-admittance of each antenna Y_1 and Y_2 and admittances characterising the coupling effect. (Reproduced by permission of © 1993 IEEE [42])

and in the case of a magnetic wall, the even admittance will be:

$$Y_{\text{even}} = Y_0 + Y_2 \tag{5.8}$$

The combination of the above three equations allows the coupling coefficient to be expressed in terms of Y_{even} and Y_{odd} only:

$$C(\text{dB}) = 20 \log \left| \frac{Y_{\text{odd0}} - Y_{\text{even}}}{Y_{\text{odd0}} + Y_{\text{even}}} \right| \tag{5.9}$$

Therefore, calculation of the even and odd admittances will yield the coupling coefficient.

As the rod has discontinuity, the scattering matrix of each discontinuity between two dielectric portions, as well as that of the dielectric portion/free space discontinuity, should be calculated as accurately as possible. The overall multimodal admittance matrices can be deduced from the multimodal scattering matrices:

$$[Y_{\text{even}}] = ([I] - [S_{\text{even}}])([I] + [S_{\text{even}}])$$

$$[Y_{\text{odd}}] = ([I] - [S_{\text{odd}}])([I] + [S_{\text{odd}}]) \tag{5.10}$$

where $[I]$ is the unity matrix, and Y_{even} and Y_{odd} will then be the first elements of the multimodal matrices $[Y_{\text{even}}]$ and $[Y_{\text{odd}}]$.

The scattering matrix of each discontinuity can then be calculated by a multimodal variation method [43], where the propagating modes on either sides of the discontinuity are known. The corresponding propagation constants can then be determined by the transverse operator method [44] by placing the antenna in an oversized waveguide. This is assumed not to disturb the main radiation pattern since the guide and the endfire antenna have the same longitudinal axis. It was observed that knowledge of the global scattering matrix of the system allows the straightforward calculation of the transverse electric field. By presenting a unit wave at the system input when both input and output are matched, the transverse electric field in the aperture is given by the following expression:

$$E_t = \sum_{n=1} t_n e_n \tag{5.11}$$

in which e_n and t_n represent the electric field and transmission coefficient of the nth mode, respectively.

The dimensions of the rod antenna are chosen such that a surface wave can propagate along the dielectric rod. Therefore it is taken that $k_z = 1.1k_0$, with the condition that the propagation constant k_z be greater than the free space wave number k_0; this guarantees the propagation of a surface wave. In contrast, $k_z/k_0 \leq 1$ should be true near the antenna's end so that a bulk wave is established in this region to ensure radiation. The number of dielectric waveguide sections is determined on the basis of the convergence of the first element of $[S_{\text{odd}}]$, and in this case, it can be shown that seven sections are sufficient to obtain this convergence.

The coupling coefficient can be calculated as a function of the distance between the longitudinal axes of the two antennas. Figure 5.19 illustrates the results obtained for the coupling coefficient at 60 GHz. These curves can be used as design curves for the dual antenna system.

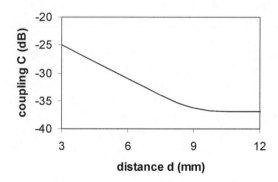

Figure 5.19 Coupling coefficient versus the interaxis distance between two rod antennas at 60 GHz

5.5 Patch-Fed Circular Rod Antennas

As in the case of optical fibres, a dielectric rod can act as a guide for electromagnetic waves. However, depending on the magnitude of the discontinuity of the dielectric constant at the boundary, a considerable amount of millimetre wave power can propagate through the surface of the rod and is radiated into free space. This radiation property is used to design the rod antenna. As shown in Figure 5.20, a rod antenna can be represented by a patch and a waveguide. The rod antenna consists of a cylindrical part and a tapered part. The rod is fed by a patch, which is energised by a microstrip line connected to a coaxial connector. The antenna configuration can be easily built and integrated with other millimetre wave functional modules or planar circuits.

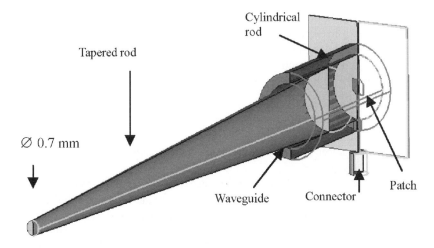

Figure 5.20   Cut-away view of a rod antenna fed by a patch and held by a waveguide. (Reproduced by permission of © 2006 IEEE [45])

A patch antenna itself produces a TM_{010} fundamental mode. When a dielectric rod is put on to the patch and is surrounded by a metallic waveguide, complex mode excitations are generated. While the antenna is radiating, energy from the patch is transferred to the tapered rod through a small cylindrical rod and a circular waveguide. The height of the cylindrical rod is set to 3 mm, while that of the circular waveguide is 7 mm. The cylindrical rod and the circular waveguide act as a mode converter, which mainly excites the TE mode. Higher modes are suppressed by selecting the appropriate height of the waveguide and diameter of the rod.

If the diameter of the cylindrical rod antenna is smaller than a quarter-wavelength, only a small amount of the energy is kept inside the rod; which also shows little guiding effect on the wave. The phase velocity in the rod is nearly the same as in free space. When the diameter increases to the order of one wavelength, most of the electromagnetic waves are held by the rod and their phase velocity in the rod is approximately the same as the phase velocity in a boundless dielectric material.

The dominant mode on the tapered rod is HE_{11} generated by a circular waveguide. The lowest mode in a circular waveguide is TE_{11} when the diameter of the guide is no less than

$0.58 \, \lambda/\sqrt{\varepsilon_r}$, where λ is the wavelength and ε_r is the relative permittivity [46]. Thus, for a rod terminated by a circular waveguide, the guide diameter must be at least 0.37λ to allow the HE_{11} mode to propagate in the metal tube.

As the dielectric constant of Teflon (PTFE) material is 2.1, the diameter of the rod antenna is designed to be 3 mm for a frequency of 61 GHz with the height 3 mm at its base, as shown in Figure 5.20. The upper part of the rod is tapered linearly to a terminal with a 0.7 mm diameter and a height of 30 mm in order to achieve a high antenna gain. The tapered rod can be treated as an impedance transformer, which reduces the reflection caused by an abrupt discontinuity [47, 48].

Teflon may not be mechanically stiff enough to enable precise manufacturing. One way to solve this issue is to freeze the Teflon rod to a low temperature in order to increase its hardness before machining.

While being fed by a patch, the diameter of the rod antenna also matches the small waveguide at its base. The inner diameter of the waveguide is 3 mm, which is the same as the diameter of the cylindrical rod. This waveguide conducts electromagnetic energy between the rod antenna and the patch antenna. This decreases the radiation leaks in unwanted directions and reduces the sidelobes in the far-field radiation pattern. In addition, this waveguide can ensure good alignment between the rod antenna and the patch antenna, which means that the design can be used in mass production. It does not even have to touch the ground plane when a thin substrate is used. The waveguide can be fixed on to the substrate by means of epoxy resin or a mechanical fixture.

The height of the cylindrical rod and the tapered rod in this example follows Zucker's design rules [49]. As a general rule, when the height of the rod antenna is reduced, the gain will reduce and the half-power beamwidth increases.

The patch antenna is designed on the Rogers RT/Duroid 5880 substrate with a dielectric constant of 2.2 and thickness of $110\,\mu m$, and so the resulting size of the 61 GHz patch is approximately 0.7 mm × 0.7 mm. The material has a similar permittivity to the Teflon rod and therefore the electric field in both materials matches reasonably well.

A rod antenna with a circular cross-section has a symmetrical shape and therefore it can generate the same energy in both right-hand circular polarisation (RHCP) and left-hand circular polarisation (LHCP). If the rod antenna is designed in an unsymmetrical shape, it will radiate more energy in one circular polarisation than the other. For instance, the top surface of the rod antenna can be designed to have an oblique upper face instead of an orthogonal surface. Assuming that the rod antenna is fed by a truncated square patch with a microstripline from the $-y$ axis to the $+y$ axis direction, different top surfaces of the rod antenna correspond to different polarisations, as shown in Figure 5.21.

The rod antenna is a directional antenna and radiates along the central axis of the rod. The direction of the main beam can be easily adjusted by changing the direction of the central axis of the rod antenna. Figure 5.22 [45] shows an example of using the rod antenna in different radiation directions. The central axis of the waveguide is also modified to fit the shape of the rod antenna.

Figure 5.23 illustrates the dependency of the radiation pattern on rod heights of 10, 18 and 30 mm. The radiation pattern at 61 GHz is simulated with the height of the tapered rod varying. As the height increases, the gain increases and the half-power beamwidth reduces. However, as the height increases to 22 mm, the antenna gain is saturated between 16.5 and 17 dBi. Distortion of the main beam occurs, which results in the presence of sidelobes at $-40°$.

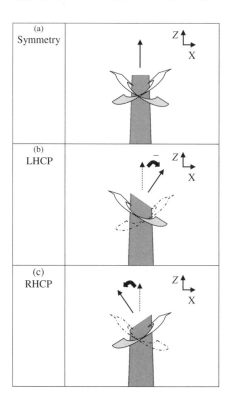

Figure 5.21 Polarisations change when the top surface of a rod antenna is altered. (Solid arrows show the rotation direction of an enhanced electrical field and dotted arrows show the rotation direction of a destructive electrical field.) (a) Symmetry, (b) LHCP and (c) RHCP. (Reproduced by permission of © 2006 IEEE [45])

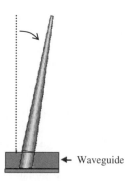

Figure 5.22 Geometry of the beam-tilting rod antennas. The beam direction changes as the rods tilt. (Reproduced by permission of © 2006 IEEE [45])

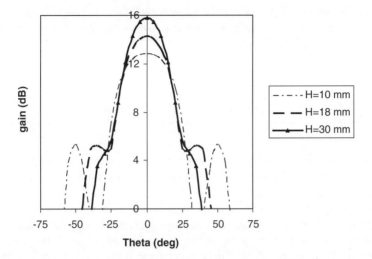

Figure 5.23 Simulated radiation pattern versus the height (H) of a tapered Teflon rod at phi $= 0$. (Reproduced by permission of © 2006 IEEE [45])

In this work, to measure the return loss, an Anritsu VP™ connector was used as an interface between the antenna and measurement equipment. The return loss of the antenna, plotted in Figure 5.24, was characterised using a vector network analyser. It can be seen that the bandwidth increases when the tapered rod is added on top of the patch antenna.

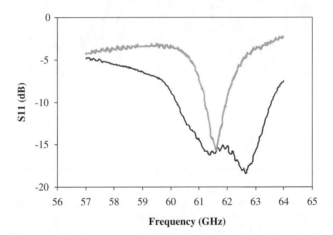

Figure 5.24 Measured S_{11} of the patch antenna with a tapered rod (black line) and without a rod (grey line). (Reproduced by permission of © 2006 IEEE [45])

Table 5.2 shows that a higher gain can be achieved by tuning the height of the waveguide and adjusting the height of the tapered rod, assuming that the height of the cylindrical rod is fixed at 3 mm. When the height of the waveguide increases from 3.5 to 7 mm, the gain

Table 5.2 Calculated antenna gain at different waveguide heights and tapered rod heights. (The height of the cylindrical rod = 3 mm.) (Reproduced by permission of © 2006 IEEE [45])

Waveguide height (mm)	Tapered rod height	24 mm	27 mm	30 mm
3.5		15.22 dBi	15.46 dBi	15.74 dBi
7		16.19 dBi	16.59 dBi	16.74 dBi

increases by approximately 1 dB. If the height of the waveguide is increased further, the gain will not be further improved as the metallic waveguide has achieved resonance and so reduces the radiation from the tapered rod.

Figure 5.25 compares simulation and measurement results of the maximum gain of a patch antenna with a rod. The patch gain increases by up to 15 dB in the frequency band of 59–65 GHz when a rod is added on top. In addition, the frequency response of the patch-fed rod in Figure 5.26 is in the range of −2 dB. This characteristic is especially useful for an extremely high rate (beyond Gb/s) communication system where a complex equaliser may be rendered unnecessary.

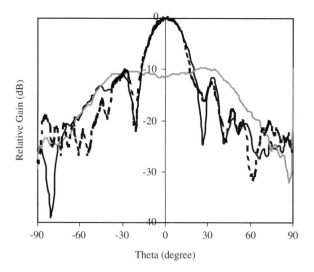

Figure 5.25 Measured E-plane radiation pattern of a patch antenna with a tapered rod at 61GHz (solid black line), 63 GHz (dotted black line) and without the rod at 61GHz (grey line). (Curves are normalised to the maximum of the rod pattern at 61 GHz.) (Reproduced by permission of © 2006 IEEE [45])

When the shape of a rod is designed asymmetrically as shown in Figure 5.21 (c) and the cutting angle Φ is set to 60°, the right-hand circular polarisation is present while the left-hand circular polarisation is suppressed.

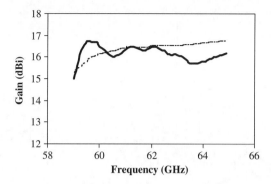

Figure 5.26 Measured (solid line) and simulated (dotted line) maximum gain of the patch-fed rod antenna. (Reproduced by permission of © 2006 IEEE [45])

The axial ratio for the patch-fed rod is measured from 59 to 63 GHz, as shown in Figure 5.27. The cross polarisation is determined by the cut-off angle and the design of the circular polarised patch. By tuning these two parameters, the cross polarisation level can be tailored to fit various applications.

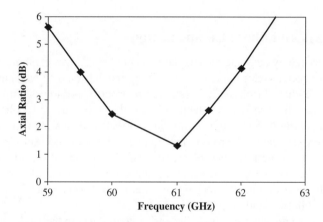

Figure 5.27 Measured axial ratio of the patch-fed rod antenna with the geometry defined in Figure 5.21 (c). (Reproduced by permission of © 2006 IEEE [45])

The beam direction can be adjusted by tilting the axis of the rod, as shown in Figure 5.22. When considering the effect of integrating this antenna with consumer devices, the ground plane is designed to be electrically large and has the size of 20 λ by 20 λ. Measurement results in Figure 5.28 show the beam at 0°, 10°, 30° and 50°; with the axis of the rod antenna is tilted to these angles. The beam direction can be seen to equal to the tilt angle of the rod. When the tilt angles increase to 50°, the asymmetrical effect of the radiation to the ground (or XY) plane becomes noticeable so the sidelobe level increases and distortion of the radiation pattern appears.

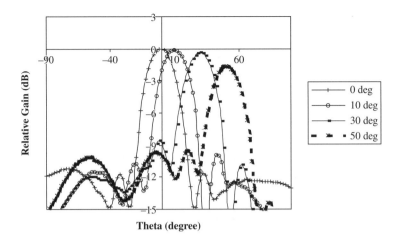

Figure 5.28   Measured normalised E-plane radiation pattern with beam direction at 0°, 10°, 30° and 50°. (All the beams are normalised to the maximum of the zero-degree beam.) (Reproduced by permission of © 2006 IEEE [45])

5.6 Rod Arrays and Phase Compensation

If multiple rods are fed by the same rectangular waveguide as shown in Figure 5.29, there will be phase delay for each rod due to the spacing of the segments on the rectangular waveguide. To make all rods radiate in phase, it is possible to insert a portion of different dielectric material into each rod base and hence alter the electrical length of the rod. If the height and the dielectric constant of this dielectric portion are carefully selected, this part acts as an impedance transformer and ensues that all the rods radiate in phase (the array is co-phased).

Figure 5.29 plots the performance of a compressed array in which all the rods radiate co-phased wavelets, and the guide feeds at an interelement spacing of less than the guide wavelength λ_g. In this example, the spacing between the rods is equal to $2\lambda_g/3$, i.e. there are three polyrods within two guide wavelengths λ_g.

If the circular waveguide is filled with dielectric material where the guide wavelength is λ_g and the cut-off wavelength is then:

$$\lambda_g = \frac{\lambda \lambda_c}{\sqrt{\lambda_c^2 - \lambda^2}} = \frac{\lambda_0 \lambda_c}{\sqrt{\varepsilon \mu \lambda_c^2 - \lambda_0^2}} \tag{5.12}$$

where λ_0 denotes the wavelength in free space, λ the wavelength in the dielectric medium, ε the electric permittivity of the medium, and μ is its magnetic permeability.

The circular waveguide is fed at its basis with a wave that has a phase retardation of $4\pi/3$ due to its propagation length L between the rods along the rectangular guide, with respect to the wave feeding the circular guide at its basis.

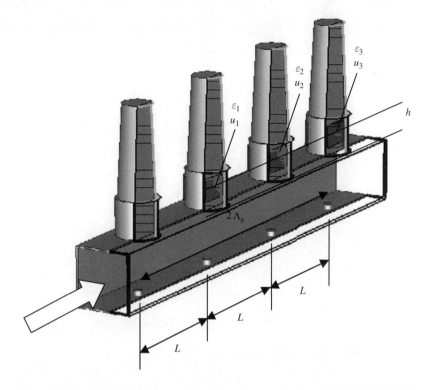

Figure 5.29 Dielectric rod array. (Reproduced by permission of © 2006 IEEE [50])

In the circular guide, a portion of rod with height h is replaced with a dielectric substance of permittivity ε_1, permeability u_1 and guide wavelength λ_{g1}, so that the phase shift delay introduced by the portion of the dielectric substance, added to the phase-shift delay is:

$$2\pi \frac{L}{\lambda_g} = \frac{4\pi}{3}$$

due to the propagation, would give a total phase shift of 2π, and with the height of the dielectric substance acting as a half-wave transformer. Then:

$$h = \frac{\lambda_{g1}}{2} = \frac{1}{2} \frac{\lambda_0 \lambda_c}{\sqrt{\varepsilon_1 \mu_1 \lambda_c^2 - \lambda_0^2}} \tag{5.13}$$

and:

$$2\pi \frac{h}{\lambda_g} = \frac{4\pi}{3} + \frac{2\pi h}{\lambda_{g1}} - 2\pi = -\frac{2\pi}{3} + \pi = \frac{\pi}{3}$$

$$\frac{h}{\lambda_g} = \frac{1}{6}$$

By substituting to h and to λ_g from their respective expressions embodied in Equations (5.12) and (5.13):

$$\frac{\sqrt{\varepsilon\mu\lambda_c^2 - \lambda_0^2}}{2\sqrt{\varepsilon_1\mu_1\lambda_c^2 - \lambda_0^2}} = \frac{1}{6}$$

$$\varepsilon_1\mu_1 = \varepsilon\mu - 8\left(\frac{\lambda_0}{\lambda_c}\right)^2$$

Assuming $\mu_1 = \mu = 1$:

$$\varepsilon_1 = \varepsilon - 8\left(\frac{\lambda_0}{\lambda_c}\right)^2$$

By choosing a wavelength in free space:

$$\lambda_0 = 3.34\,\text{mm}$$

and a circular guide of a radius of 1.6 mm, and therefore a cut-off wavelength of:

$$\lambda_c = 1.6 \times 1.7 = 2.72\,\text{mm}$$

corresponding to a wave H_{11}, and by taking:

$$\varepsilon = 2.25$$

therefore:

$$\varepsilon_1 = 9 \times 2.25 - 8\left(\frac{3.34}{2.72}\right)^2 = 8.15$$

$$h = 0.65\,\text{mm}$$

In the same manner, in the circular guide, a portion of a dielectric substance will be inserted with height h, permittivity ε_2 and permeability μ_2 and a guide wavelength λ_{g2} so that the phase-shift delay introduced by this portion of the dielectric substance, added to the phase-shift delay is:

$$2\pi\frac{2L}{\lambda_2} = \frac{8\pi}{3}$$

due to the propagation, would give a total phase shift of 4π, and the height of the dielectric substance will act as a three-halves wave transformer. Then:

$$h = \frac{3\lambda_{g2}}{2} = \frac{3\lambda_c\lambda_0}{2\sqrt{\varepsilon_2\mu_2\lambda_c^2 - \lambda_0^2}} \tag{5.14}$$

and:

$$2\pi\frac{h}{\lambda_g} = \frac{8\pi}{3} + \frac{2\pi h}{\lambda_{g2}} - 4\pi = -\frac{4\pi}{3} + 3\pi = \frac{5\pi}{3}$$

By substituting to h and to λ_g from their respective expressions embodied in Equations (5.13) and (5.14):

$$\frac{3\sqrt{\varepsilon\mu\lambda_c^2 - \lambda_0^2}}{2\sqrt{\varepsilon_2\mu_2\lambda_c^2 - \lambda_0^2}} = \frac{5}{6} \tag{5.15}$$

$$\varepsilon_2\mu_2 = \frac{81}{25}\varepsilon\mu - \frac{56}{25}\left(\frac{\lambda_0}{\lambda_c}\right)^2 \tag{5.16}$$

Assuming $\mu_2 = \mu = 1$:

$$\varepsilon_2 = \frac{81}{25}\varepsilon - \frac{56}{25}\left(\frac{\lambda_0}{\lambda_c}\right)^2 \tag{5.17}$$

Adopting for λ_0, λ_c and ε the same values as before, therefore:

$$\varepsilon_2 = 3.96$$

Using the same principle, ε_3 can also be calculated.

5.7 Optimisation of a Rod Antenna

In this section a basic physical model of how the rod works is presented. An open-ended circular waveguide supporting the dominant TE_{11} mode is often used as a feed for paraboloidal reflectors. The aperture diameter is usually close to $0.7\,\lambda_0$ when the E- and H-plane radiation patterns are approximately equal and follow a $\cos\theta$ law (where λ_0 is the free space wavelength and θ is the angle off-axis). As a feed, the radiated power in the far field can be considered as a point source in the centre of the aperture.

It is assumed that such a point source exists in the aperture plane of the polyrod launcher with the same directional properties. In addition, ray optics can be applied within the rod, and rays will either be refracted or be totally internally reflected, depending on their incident angle at the interface between the dielectric and the air.

In Figure 5.30, four rays (i), (ii), (iii) and (iv), are shown leaving the source, two of which, (i) and (ii), leave the rod close to the launcher and (iii) and (iv) traverse the height of the rod by total internal reflection. Rays (i) and (ii) are associated with a short section of the rod at the launcher end where the angle of incidence is less than critical, and ray (iii) emerges from abrupt termination with ray (iv) and is reflected back towards the source.

As can be seen in a tapered rod, rays will increase their angle to the axis by twice the local taper angle at each reflection. Eventually, given a sufficient number of reflections, a ray will reach the critical angle and escape from the rod. In the case of a linear taper to the terminal after a cylindrical section, as the percentage of taper increases the radiating rays will be distributed over the height of the taper. The rod will therefore appear to be comprised of a constant amplitude launcher end source, a zero amplitude section to the beginning of the taper, a distributed source over the taper and a diminishing amplitude terminal end source, with increasing taper height.

The power distribution along the rod affects sidelobe performance. It is therefore necessary to have some form of continuously tapering profile so that gaps in the distribution are avoided.

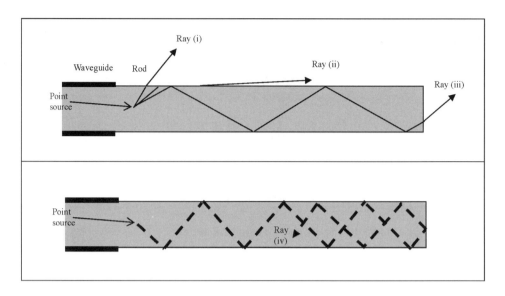

Figure 5.30   Outward and inward ray paths in a cylindrical rod. (Reproduced by permission of © 2006 IEEE [51])

Also, the weaker rays associated with wide angles from the directional source are to be encouraged to exit the rod as soon as possible if a balanced distribution is to be obtained. This can be done by increasing the slope of the taper near the launcher end, consistent with the appropriate diameter for correct phasing.

The model can be used to predict the rod diameter around the launcher; where internal reflections take place near the critical angle and phase change at the reflection can be disregarded. Figure 5.31 shows two rays: one incident on the dielectric-to-air interface at the critical angle θ_c leaving parallel to the rod surface, and the other leaving the rod after two reflections. The electrical lengths of the two ray paths in Figure 5.31 for endfire operation can be written as:

$$L_0\sqrt{\varepsilon} + \text{air path} + n\lambda_0 = \sqrt{\varepsilon}(L_1 + L_2 + L_3) \tag{5.18}$$

where $n = 1, 2, 3, \ldots$ and ε is the dielectric constant of the rod.

The model improves the match obtained by including a terminal taper such as a short height at the end of the rod, which gives a large taper angle. In this section rays still trapped by total internal reflection rapidly achieve the critical angle and escape, so reducing the number that can return as mismatch.

Therefore, it can be seen that for a smooth power distribution along the rod, a continuously tapering profile is necessary. Whether the point source characteristics can be combined with a profile, that gives a balanced distribution of power and phase is a topic for future research. However, by measuring rods that have known profile laws and by determining the position of the phase centre, an idea of the feasibility of controlling the distribution can be obtained. The phase centre is found to be at or near the rod centre, indicative of a balanced distribution.

To reduce unwanted radiation at the launcher end and to make more power available for controlled distribution, the model shows that high dielectric constant rods are preferable.

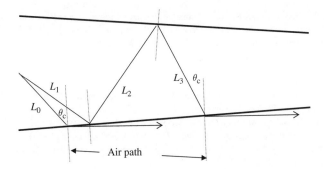

Figure 5.31   Geometry of two rays close to the critical angle [51]

Nevertheless, the loss tangent of the material plays a key role at millimetre wave frequency and has to be considered in real applications.

Designs based on these models shows that the gain can be regarded as a minimum of 16 dBi, for a gain times beamwidths product of approximately 27 000. Using the model as a design method, further improvements may result from investigations into polyrods made from low-loss, high-dielectric constant materials.

References

[1] Satoshi Kobayashi, Raj Mi'ptra and Ross Lampe, 'Dielectric Tapered Rod Antennas for Millimetre-Wave Applications', *IEEE Transactions on Antennas and Propagation*, **AP-30**(1), January 1982, 54–58.

[2] Kwai-Man Luk and Kwok-Wa Leung, *'Dielectric Resonators'*, Artech House, Massachusetts, 1986.

[3] D. Hondros, 'Ueber elektromagnetische Drahtwelle', *Annalen der Physik*, **30**, 1909, 905–949.

[4] H. Zahn, ,Ueber den Nachweis elektromagnetischer Wellen an dielektrischen Draehten', *Annalen der Physik*, **37**, 1916, 907–933.

[5] S. A. Long, M. W. McAllister and L. C. Shen, 'The Resonant Cylindrical Dielectric Cavity Antenna', *IEEE Transactions on Antennas and Propagation*, **31**, May 1983, 406–412.

[6] R. K. Mongia and A. Ittipiboon, 'Theoretical and Experimental Investigations on Rectangular Dielectric Resonator Antennas', *IEEE Transactions on Antennas and Propagation*, **45**, September 1997, 1348–1356.

[7] R. K. Mongia, A. Ittipiboon, P. Bhartia and M. Cuhaci, 'Electric Monopole Antenna Using a Dielectric Ring Resonator', *Electronics Letters*, **29**, August 1993, 1530–1531.

[8] Countis Laboratories, http://countis.com/

[9] ECCOSTOCK, http://www.eccosorb.com/

[10] S. A. Long, M. McAllister and L. C. Shen, 'The Resonant Cylindrical Dielectric Cavity Antenna', *IEEE Transactions on Antennas and Propagation*, **AP-31**, May 1983, 406–412.

[11] A. W. Glisson, D. Kajfez and J. James, 'Evaluation of Modes in Dielectric Resonators Using a Surface Integral Equation Formulation', *IEEE Transactions on Microwave Theory and Techniques*, **MTT-31**, December 1983, 1023–1029.

[12] R. K. Mongia and P. Bhartia, 'Dielectric Resonator Antennas – A Review and General Design relations for Resonant Frequency and Bandwidth', *International Journal of Microwave and Millimetre-Wave Computer-Aided Engineering*, **4**(3), 1994, 230–247.

[13] A. Petosa, A. Ittipibon, Y. M. M. Antar, D. Roscoe and M. Cuhaci, 'Recent Advances in Dielectric Resonator Antenna Technology', *IEEE Antennas and Propagation Magazine*, **40**(3), June 1998, 35–48.

[14] Kwai-Man Luk and Kwok-Wa Leung, *'Dielectric Resonators'*, Artech House, Norwood, Massachusetts, 1986, p. 197.

[15] A. A. Kishk, A.W. Glisson and D. Kajfez, 'Computed Resonant Frequency and Far Fields of Isolated Dielectric Discs', *IEEE Antennas and Propagation Society International Symposium Digest*, **1**, 1993, 408–411.

[16] L. B. Felsen, 'Radiation from a Tapered Surface Wave Antenna', *IRE Transactions on Antennas and Propagation*, **AP-8**, November 1960, 577–586.

[17] F. J. Zucker, 'Surface and Leaky-Wave Antennas', in *'Antenna Engineering Handbook'*, Ed. H. Jasik, McGraw-Hill, New York, 1961, Chapter 16.

[18] Y. Shiau, 'Dielectric Rod Antennas for Millimetre-Wave Integrated Circuits', *IEEE Transactions on Microwave Theory and Techniques*, **MTT-24**, November 1976, 869–872.

[19] R. E. Collin and F. J. Zucker, *'Antenna Theory'*, McGraw-Hill, New York,1969, Part 2.

[20] J. B. Andersen, *'Metallic and Dielectric Antennas'*, Polyteknisk Forlag, Denmark, 1971.

[21] J. R. James, 'Engineering Approach to the Design of Tapered Dielectric-Rod and Horn Antennas', *Radio Electronics Engineering*, **42**(6), 1972, 251–259.

[22] F. J. Zucker, 'Surface-Wave Antennas and Surface-Wave Excited Arrays', in *'Antenna Engineering Handbook'*, 2nd edition, Eds R. C. Johnson and H. Jasik, McGraw-Hill, New York, 1984, Chapter 12.

[23] Y. Shiau, 'Dielectric Rod Antennas for Millimetre-Wave Integrated Circuits', *IEEE Transactions on Microwave Theory and Techniques*, **MTT-24**, November 1976, 869–872.

[24] T. Takano and Y. Yamada, 'The Relation between the Structure and the Characteristics of a Dielectric Focused Horn', *Transactions of IECE*, **J60-B**(8), 1977, 395–593.

[25] S. Kobayashi, R. Mittra and R. Lampe, 'Dielectric Tapered Rod Antennas for Millimetre-Wave Applications', IEEE Transactions on Antennas and Propagation, **30**, January 1982, 54–58.

[26] C. Yao and S. E. Schwarz, 'Monolithic Integration of a Dielectric Millimetre-Wave Antenna and Mixer Diode: an Embryonic Millimetre-Wave IC', *IEEE Transactions on Microwave Theory and Techniques*, **MTT-30**, August 1982, 1241–1247.

[27] R. Chatterjee, *'Dielectric and Dielectric-Loaded Antennas'*, Research Studies Press, UK, 1985.

[28] F. Schwering and A. A. Oliner, 'Millimetre-Wave Antennas', in *'Antenna Handbook'*, Eds Y. T. Lo and S. W. Lee, Van Nostrand Reinhold, New York, 1988, Chapter 17.

[29] M. Koshiba, *'Optical Waveguide Analysis'*, McGraw-Hill, New York, 1990, Chapter 5.

[30] C. Vassallo, '1993–1995 Optical Mode Solvers', *Optical Quantum Electronics*, **29**, 1997, 95–114.

[31] D. Yevick and W. Bardyszewski, 'Correspondence of Variation Finite Difference (Relaxation) and Imaginary-Distance Propagation Methods for Modal Analysis', *Optical Letters*, **17**(5), 1992, 329–330.

[32] S. Xiao, R. Vahldieck and H. Jin, 'Full-Wave Analysis of Guided Wave Structure Using a Novel 2-D FDTD', *IEEE Microwave Guided Wave Letters*, **2**, 1992, 165–167.

[33] A. Asi and L. Shafai, 'Dispersion Analysis of Anisotropic Inhomogeneous Waveguides Using Compact 2D-FDTD', *Electronics Letters*, **28**, 1992, 1451–1452.

[34] S. M. Lee, 'Finite-Difference Vectorial-Beam-Propagation Method Using Yee's Discretization Scheme for Modal Fields', *Journal of Optical Society of America A, Optical Image Science*, **13**(7), 1996, 1369–1377.

[35] J. Yamauchi, N. Morohashi and H. Nakano, 'Rib Waveguide Analysis by the Imaginary-Distance Beam-Propagation Method Based on Yee's Mesh', *Optical Quantum Electronics*, **30**, 1998, 397–401.

[36] T. Ando, J. Yamauchi and H. Nakano, 'Demonstration of the Discontinuity- Radiation Concept for a Dielectric Rod Antenna', *Proceedings of IEEE AP-S International Symposium Digest*, 2000, 856–859.

[37] T. Ando, J. Yamauchi and H. Nakano, 'Numerical Analysis of a Dielectric Rod Antenna', *IEEE Transactions on Antennas and Propagation*, **51**(8), August 2003, 2007–2013.

[38] S. T. Chu, W. P. Huang and S. K. Chaudhuri, 'Simulation and Analysis of Waveguide Based Optical Integrated Circuits', *Computational Physics Communications*, **68**, 1991, 451–484.

[39] A. Taflove and S. C. Hagness, *'Computational Electrodynamics, The Finite- Difference Time-Domain Method'*, 2nd edition, Artech House, Norwood, Massachusetts, 2000.

[40] O. M. Ramahi, 'The Concurrent Complementary Operators Method for FDTD Mesh Truncation', *IEEE Transactions on Antennas and Propagation*, **46**, October 1998, 1475–1482.

[41] T. N. Trinh, J. A. Malberk and R. Mittra, 'A Metal-to-Dielectric Waveguide Transition with Application to Millimetre-Wave Integrated Circuits', IEEE MTT-S International Microwave Symposium, May 1980, pp. 205–207.

[42] M. Aubrion, A. Larminat *et al.*, 'Design of a Dual Dielectric Rod-Antenna System', *IEEE Microwave and Guided Wave Letters*, **3**(8), August 1993, 276–280.

[43] J. W. Tao and H. Baudrand, 'Multimodal Variational Analysis of Uniaxial Waveguide Discontinuities', *IEEE Transactions on Microwave Theory and Techniques*, **39**, March 1992, 1–11.

[44] J.W. Tao, J. Atechian, R. Ratovondrahanta and H. Baudrand, 'Transverse Operator Study of a Large Class of Multidielectric Waveguides', *IEE Proceedings*, **137**, October 1990, 135–139.

[45] K. Huang and Z. Wang 'V-Band Patch-Fed Rod Antennas for High Data-Rate Wireless Communications', *IEEE Transactions on Antennas and Propagation*, **54**(1), January 2006, 297–300.

[46] J. Kraus and R. Marhefka, *'Antennas for All Applications'*, 3rd edition, McGraw-Hill, New York, 2002.

[47] S. Kobayashi, R. Mittra and R. Lampe, 'Dielectric Tapered Rod Antenna for Millimetre Wave Applications', *IEEE Transactions on Antennas and Propagation*, **30**(1), January 1982, 54–58.

[48] Y. Shiau, 'Dielectric Rod Antenna for Millimetre Wave Integrated Circuits', *IEEE Transactions on Microwave Theory and Techniques*, **24**(11), November 1976, 869–872.

[49] F. J. Zucker, *'Antenna Engineering Handbook – Surface and Leaky-Wave Antenna'*, 3rd edition, McGraw-Hill, New York, 1992.

[50] Maurice G. Bouix, 'Dielectric Antenna Array', US Patent 2624002.

[51] A.C. Studd, 'Towards a Better Dielectric Rod Antenna', IEE Seventh International Conference on *'Antennas and Propagation'*, ICAP 91, Vol. 1, 15–18 April 1991, pp. 117–120.

6

Lens Antennas

In the current context, lenses are made from low-loss dielectric materials that have a higher dielectric constant than air. In general, at millimetre wave frequencies quasi-optical principles can be applied to this type of antenna, as it works in a similar manner to electromagnetic waves at millimetre wave frequencies. In optical terminology, the index of refraction n is used in place of the dielectric constant ε_r:

$$n = \sqrt{\varepsilon_r}$$

The wave impedance of the material can be obtained from the index of refraction [1]:

$$Z = \sqrt{\frac{\mu}{\varepsilon}} = \frac{\eta_0}{n} = \frac{377}{n}$$

where the impedance of free space is taken as 377 Ω.

When waves encounter an impedance discontinuity they are partially transmitted and partially reflected in a similar manner to transmission line impedance mismatches. Also, when the radiation is incident at an angle other than perpendicular to the surface, refraction or bending of light electromagnetic radiation occurs (Snell's law). The "rays" are bent towards the surface normal when entering a medium with a higher dielectric constant, and towards the surface normal when going from a higher to a lower dielectric constant material. (This is generally known as geometric optics.) This approach is fairly accurate for structures that are large compared to the wavelength of the radiation.

Thus, when emerging from the second surface of a converging lens (which has a shaped surface), the rays are bent once again. The lens is constructed with a curvature such that rays incident at different points are bent by a different amount. Therefore, it can be seen that the lens acts as a concentrator, gathering energy over an area and concentrating it to a point.

For those unfamiliar with basic optics a brief description of the major features lenses will now be given. Consider a convex (converging) lens where the source is a long way away (ideally at infinity) so that the incident wavefront is planar, and with the incident rays perpendicular to the major axis of the lens, as shown in Figure 6.1.

Millimetre Wave Antennas for Gigabit Wireless Communications   Kao-Cheng Huang and David J. Edwards
© 2008 John Wiley & Sons, Ltd

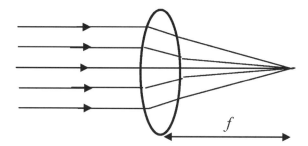

Figure 6.1 A converging lens focusing light from an infinite distance. The light converges at the focal length, f, of the lens

The ray that hits the centre of the lens is exactly orthogonal to both of the lens surfaces at this point, and therefore its direction is not changed. Rays that hit the lens at points off-centre form an angle with the local surface normal. These rays will therefore be refracted. As can be seen, the incident angle and thus the refraction angle changes with distance from the centre (axis) of the lens. Rays at the edge of the lens will be bent the most. The result is that all the rays are focused at a single point behind the lens. This point is called the focal point of the lens and the distance from the lens to that point is known as the focal length. This description is sufficient for thin lenses and higher-order aberration effects will not be considered at the moment.

Another interesting point is that the waves that follow each ray have exactly the same path length (Fermat's principle), and therefore arrive "synchronised" or in phase. The rays may have different physical lengths, but the slower speed of light inside the lens ($v = c/n$) causes a delay. For example, the ray that hits the centre point of the lens is delayed the most because it travels through the thickest part of the lens. The rays that go through the lens near the edge are delayed the least because they travel through the thinnest part of the lens. These rays, however, travel the longest distance through air. Thus the delay through air plus the delay through the lens is always the same for rays that arrive at the focal point. Therefore, these characteristics can be applied to the development of lens antennas.

6.1 Luneberg Lens

When a multibeam lens employs an array of primary feed elements, the high number of primary feed ports requires additional inputs or outputs in the beam selection switch. This of course causes an increase in attenuation and can result in a loss of more than 1 dB per switch. A different concept for this multibeam antenna approach was proposed by R. K. Luneberg in 1943 [2]. He proposed the principle of this lens for electromagnetic waves. The general principle is that a sphere made of materials with a relative dielectric constant ε_r varies as the square of the distance from the surface to the centre, and becomes a dielectric lens with foci lying on a surface:

$$\varepsilon_r = 2 - (r/R)^2$$

where ε_r is relative dielectric constant, r is radius from the centre point and R is the outer radius of the lens.

This suggests that a single-lens antenna of this type is capable of receiving and transmitting waves from, and in multiple directions, at the same time, providing that multiple feed elements are provided at the focal surface. Luneberg did not have the opportunity to implement such an antenna, as no suitable materials or manufacturing procedures were available at that time.

Radio waves refract, just like light, at the interface of two materials with different relative dielectric constants. An electromagnetic wave entering at a focus on the surface is refracted at each interface within the dielectric material sphere, as the relative dielectric constant changes gradually from the surface to the centre, and is eventually emitted as a plane wave from the opposite side of the sphere.

A hemispherical lens antenna is commonly used because of the difficulty of stabilising a heavy sphere. Equivalent antenna characteristics can be obtained by placing a flat reflector on the base. The design is illustrated in Figure 6.2.

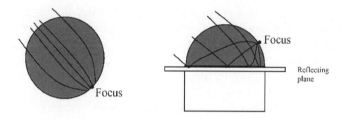

Figure 6.2   Luneberg lens and hemispherical lens [3]

This configuration offers a relatively low-profile solution, and this property makes the hemi-sphere antenna particularly advantageous as a scanning antenna for network access points. An example application is that of an antenna mounted on the ceiling of a railway coach which is used to backhaul, via a satellite link, a wireless local area network used by passengers. It is important to consider the layout as illustrated in Figure 6.3, where it can be seen that the effective aperture height of the hemisphere with a reflecting plane can be up to twice that of a conventional reflector antenna [4].

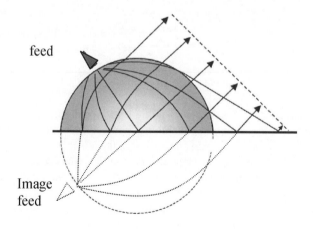

Figure 6.3   Hemisphere lens antenna

Figure 6.3 also illustrates the approximate ray paths for the Luneburg lens case and hence shows curved paths within the dielectric. For a layered structure comprising discrete shells the rays would of course describe a series of straight lines within each shell. In Figure 6.3, the aperture blockage by the feed could be avoided by tilting the ground plane and offsetting the feed by a similar angle.

In a hemispherical lens with a ground plane, it should be realised that the electric field comprises of that component which arises directly from the feed and lens, and that which is reflected from the ground plane. The latter term contributes significantly to the main lobe of a spherical lens while the former term constitutes a relatively small component. A continuous radial variation in dielectric constant is difficult to achieve in practice and so lenses of this type are usually constructed from a series of concentric shells; several design approaches have been reported [4, 5].

A simplified version of a Luneberg lens comprising just two concentric layers, as shown in Figure 6.4, can yield a useful improvement in aperture efficiency. A two-shell lens with feed fixed at 8.2λ is also worth considering, where the outer radius r_2 can be fixed at 8λ and r_1 allowed to vary. Much work has been done to achieve the design closest matching Luneburg's equation, and attempting to synthesise the necessary dielectric materials for each layer (e.g. see Reference [5]). Now the use of readily available materials that have well-characterised dielectric constants and, most importantly, low-loss tangents needs to be considered. This design approach is reported in Reference [6] where the properties of single-layer and two-layer lenses were investigated.

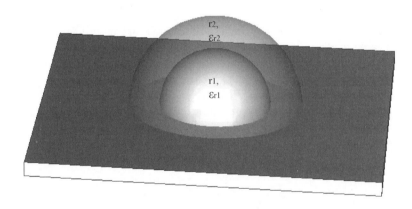

Figure 6.4 Two-layer lens geometry [3]

Different lens materials exhibit different optimum feed positions. It is observed that fused silica, having a dielectric constant of 3.8, exhibits a paraxial focus point very close to the lens outer edge (for $\varepsilon_{r1} > 4$ the paraxial focus moves inside the lens outer radius [7]) and hence the directivity reduces with increasing feed displacement in the radial direction.

There is no advantage in adding an outer layer to a fused silica core and only minimal advantage in using low-dielectric constant foam as an outer layer, as this layer would need to be very thin ($\sim \lambda$) to be beneficial, and hence difficult to fabricate. The best results were obtained using a Rexolite inner core of radius 4.2λ and a polyethylene outer core, which realised 76 % aperture efficiency. The optimal design can offer up to 36.0 dBi theoretical directivity [3].

6.2 Hemispherical Lens

The hemispherical lens has a higher mechanical stability than the spherical lens. To fabricate two-dimensional arrays using MMIC techniques, it is possible to have a lens-coupled patch antenna configuration, as shown in Figure 6.5, which is very amenable to fabrication. It is compact and has the advantage of being able to include additional integrated circuits. This is particularly important in fabricating two-dimensional arrays [8, 9].

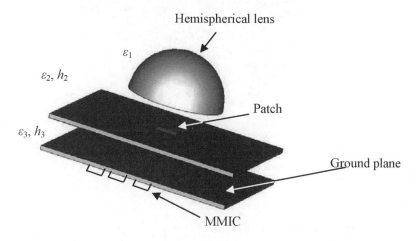

Figure 6.5 The configuration of a microstrip patch antenna

This array consists of two individual microstrip substrates separated by a (common) metal ground plane. The antennas are printed on the first substrate (ε_{r2}) covered with the low-loss dielectric lens (ε_{r1}). Each antenna is fed with a coupling slot [10, 11] from the MMIC constructed on the lower substrate (ε_3). The lower substrate offers an efficient space for fabricating additional integrated circuits such as matching circuits, mixers, amplifiers and interconnections. The antennas are isolated from these circuits by the ground plane.

Figure 6.6 shows the calculated radiation patterns of each individual antenna element in Figure 6.5 [12]. The patch is separated from MMICs by two substrates and its size is subject to the relative permittivity of the upper and lower substrates. Defining:

$$\varepsilon_{12} \equiv \frac{\varepsilon_2}{\varepsilon_1} \tag{6.1}$$

as the ratio of the dielectric constant of the first substrate ε_2 to the dielectric constant of the lens ε_1, both the patch length a and the patch width b are:

$$a = b = \frac{\lambda_{\text{eff}}}{2} \tag{6.2}$$

where the effective wavelength in the first substrate λ_{eff} is defined by:

$$\lambda_{\text{eff}} = \frac{\lambda_0}{\sqrt{\varepsilon_{\text{eff}}}} \tag{6.3}$$

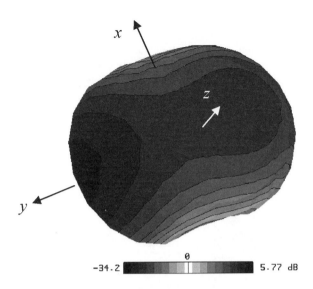

Figure 6.6 The calculated radiation patterns of the microstrip patch antenna with a dielectric hemisphere in Figure 6.5

and the effective dielectric constant of the first substrate ε_{eff} is given by [13]:

$$\varepsilon_{\text{eff}} = \frac{\varepsilon_{12} + 1}{2} + \frac{\varepsilon_{12} - 1}{2} \left(1 + \frac{10h_2}{b} \right)^{-1/2} \tag{6.4}$$

where h_2 is the thickness of the first substrate and b is the patch width [13]. If ε_{r1} equals ε_{r2} then the ratio ε_{12} becomes one, an ideal radiation pattern that is almost symmetrical for both the E- and H-planes can be realised. This pattern has neither sidelobes nor radiation in the horizontal directions, and so offers low crosstalk and high beam coupling efficiency to the incident beam. When ε_{12} does not equal one, an undesirable substrate mode is generated in the first substrate, which affects the radiation and the impedance characteristics of the adjacent antennas in the array [14].

The next consideration is related to antenna mounting. Different methods exist for mounting a lens to a planar antenna. The classical way for mounting a lens is to use mechanical holders [15]. However, these holders are heavy, expensive and introduce additional reflections, which result in changes in the radiation pattern. One possible solution is to add a foam sandwich layer between the lens and the patch itself. Since the foam has a dielectric constant close to the air, it should not influence the performance of the antenna system. However, the multiple glue layers, which had to be applied, influenced the performance of the lens at millimetre wave frequencies. To overcome these problems, an "eggcup" type of lens can be constructed, which has a small size and light weight. It is also easy to manufacture and therefore has a low cost [16].

The cross-sectional view of the lens is shown in Figure 6.7. It consists of a quasi-lens, a waveguide and a cavity. The lens can be designed using the geometric optics method. The cavity and waveguide, also work as a lens supporter. To minimise the effect to the lens and the patch, both the cavity and waveguide should have a thin dielectric wall. These are all made from the same dielectric material to maintain good impedance matching.

The cavity is designed to contain the resonant energy while the waveguide is designed to transform and filter the required mode that passes through it (Figure 6.7). A further increase in the cylinder diameter leads to multiple reflections within the cylinder and thus the performance is degraded. For ease of manufacturing, a deviation is acceptable, from the calculated lens contour within the cylindrical holder, by employing a flat surface which gives negligible degradation in the performance (x point in Figure 6.7).

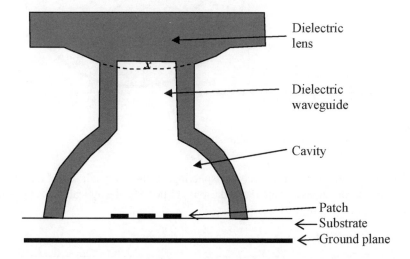

Figure 6.7 Cross-sectional view of the lens. The dielectric structure consists of a parabolic type of lens and waveguide with a cavity as the lens supporter. (Reproduced by permission of © 2006 IEEE [16])

The achieved half-power beamwidths for this arrangement are 20° in both the E-plane and the H-plane. The first sidelobes are below 15 dB. The measured and simulated gain of the complete antenna is around 15 dB over the frequency range from 57 up to 63 GHz [15]. It should be noted that dispersion losses and tolerances of the dielectric constant of the materials may affect antenna performance at the end.

6.3 Extended Hemispherical Lens

An extended hemi-spherical lens is like a semi-elliptical lens and can focus a plane wave to a point. Its principle is based on refraction at spherical surfaces (Figure 6.8). In physical optics, only the tangential electric and magnetic fields at the lens interface between the dielectric and free space are calculated. The Schelkunoff equivalence principle [17] is then applied to substitute equivalent magnetic and electric currents for the surface magnetic and electric fields, respectively, and the radiation patterns are then computed from these equivalent currents. This is also known in optics as Babinet's principle [17].

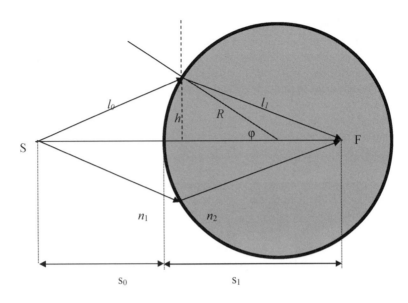

Figure 6.8 Refraction at a spherical surface

A ray from an axial point S intersects the spherical surface at height h in Figure 6.8. After refraction, the ray converges and intersects the axis at a point F. The optical path length between points S and F can be expressed as:

$$n_1 l_0 + n_2 l_1$$
$$= n_1 \sqrt{R^2 + (s_0 + R)^2 - 2R(s_0 + R)\cos\phi} + n_2 \sqrt{R^2 + (s_1 - R)^2 - 2R(s_1 - R)\cos\phi}$$

where n_1 is the index of the air and n_2 is the index of the medium. Using Fermat's principle and paraxial approximation, the refraction at spherical surfaces is as follows:

$$\frac{n_1}{s_0} + \frac{n_2}{s_1} = \frac{n_2 - n_1}{R}$$

If point S is located at a position where $s_0 >> s_1$ (e.g. plane wave), the above equation can be simplified to:

$$\frac{n_2}{s_1} \approx \frac{n_2 - n_1}{R} \text{ or } \frac{R}{s_1} + \frac{n_1}{n_2} \approx 1$$

When R and n_1 are fixed, it is found that the lens with the higher index n_2 (or higher dielectric constant) will have a smaller converging length s_1. This can be seen in Figure 6.9, showing three extended hemispherical lenses with $\varepsilon_r = 2, 4$ and 12. The higher the permittivity, the smaller is the antenna size. They can be implemented using a planar wafer. Extended hemispherical lenses can be synthesised with an ellipse. It is shown in Reference [18] that the synthesised ellipse presented better results for less than a 6 % decrease in the Gaussian coupling efficiency at 500 GHz for a 6.8 mm silicon or quartz lens from a true elliptical lens.

The focal point can also be considered to be located at the second focus of the ellipse. The shape of the extended hemispherical lens depends on the index of refraction of the lens used,

Figure 6.9   An elliptical lens superimposed on an extended hemispherical lens for different permittivities

and it is straightforward to derive the formula. The extended hemispherical lens has infinite magnification since a spherically diverging beam from the focal point is transformed into a plane wave. In antenna terms, this means any antenna placed at the focus of the extended hemispherical lens will generate a far-field pattern, with a main beam that is diffraction limited by the aperture of the extended hemispherical lens. The difference between these antennas then, is in the sidelobe and cross polarisation levels. Since the patterns are diffraction limited by the lens and therefore are very narrow, any increase in the sidelobe level can have a detrimental effect on the overall efficiency of the system. The extended hemispherical lens is compatible with small aperture imaging systems owing to its narrow diffraction-limited patterns and should be placed near the minimum waist plane, where no phase errors are present in the Gaussian beam. This is in contrast to the hyper-hemi-spherical lens, which should be placed in a converging beam (i.e. with an appropriate phase error) for maximum coupling to an optical system.

When parallel rays entering a lens do not come to focus at a point, it is said that the lens has an aberration. As can be seen in Figure 6.10, if light enters too large a region of a spherical surface, the focal points are spread out at the back. This is called spherical aberration. One solution for spherical aberration is to make sure that the diameter of any spherical lens is small in comparison to the radius of curvature of the lens surface.

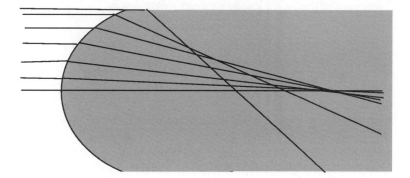

Figure 6.10   Spherical aberration

For ease of fabrication of an extended hemi-spherical lens, the dimensions are chosen to approximate to the desired focusing properties of an elliptical lens with a feed located at one of its foci:

$$b = R\sqrt{\varepsilon_r(\varepsilon_r - 1)}$$

$$d = b\sqrt{\varepsilon_r}$$

$$L = d + b - R$$

where R is the lens radius at the maximum waist, b is the major semi-axis of elliptical curvature, d is the cylindrical extension length for an elliptical lens and L is the total combined cylindrical extension for the extended hemispherical approximation.

Figure 6.11 shows a cross-sectional view of the proximity-coupled microstrip patch-lens configuration. Here a microstrip line parasitically excites a rectangular microstrip patch. Above the patch is a dielectric lens terminating a cylindrical cross-section of length L and radius $R(R = a)$(an extended lens). The lens is fabricated as an ellipsoidal lens to maximise the directivity [19]. As mentioned previously, to ensure that no power is lost to surface waves, the dielectric constant of the grounded substrate for the feed line, (the layer on which the microstrip patch is etched and the material for the lens) must have the same permittivity. As a further option, consideration can be given to a proximity-coupled patch antenna on an extended hemispherical dielectric lens for millimetre wave applications (Figure 6.11). This configuration has several advantages over the conventional microstrip antenna lens arrangements.

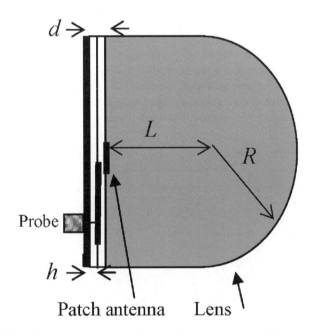

Figure 6.11   Cross-sectional view of the proximity-coupled patch lens antenna [22]

Firstly, no surface wave losses will be found associated with the feed network if the same dielectric constant materials are used for the multilayered patch configuration and the lens (as opposed to the aperture-coupled configurations in References [20] and [21]). Using a proximity-coupled patch configuration yields greater bandwidths than a direct contact fed patch lens without degrading the front-to-back ratio of the antenna, unlike the case of aperture-coupled patches [20, 21] or printed slot versions [19–28]. Non-contact feeding techniques, such as proximity coupling, also tend to have lower cross polarisation levels than direct contact excitation methods [29].

As a further point, low-cost, low-dielectric-constant materials, such as polyethylene, can be used without degradation of the front-to-back ratio, unlike a slot configuration [21, 25–28]. In the case of the proximity coupled patch antenna, this immunity to parasitic radiation is compromised [19].

The size of a lens should ensure that the lens surface is located in the far field of the printed feed radiation pattern for both the "first-order" rays, which have a single point of intercept with the lens surface as shown in Figure 6.12, and the internally reflected rays, called "second-order" rays. A lens with a radius of 12.5 mm and a permittivity of 12.0 can be used as a starting point for a 60 GHz scale model [22].

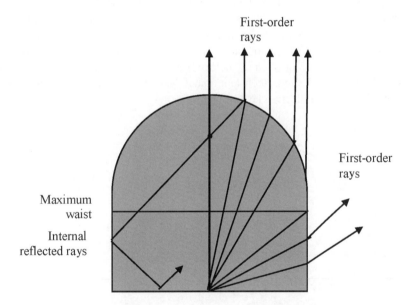

Figure 6.12 Dielectric lens modelling: two-dimensional ray tracing for an elliptical lens, with second-order internally reflected rays. (Reproduced by permission of © 2006 IEEE [30])

As can be seen in Figure 6.12, the lens collimating property is only effective over its convex surface above the plane of its maximum waist. Geometric optics analysis reveals that feed radiation intercepting the lens surface below the maximum waist, at the surface of a cylindrical extension, is not collimated, but rather propagates laterally in undesired directions. For this reason, the most efficient feed architectures for use with lens antennas should be designed to minimise radiation in lateral directions along the ground plane. Such lateral radiation can also be found for other feed architectures, such as the conventional dual-slot feed [23] and the twin arc-slot design [24].

The behaviour modelling of this antenna can be simplified by assuming that the radius and length of the extended lens are significantly greater than the dimensions of the patch antenna. Doing so allows the microstrip antenna to be represented as if it were mounted in an infinite half-space of dielectric constant ε_r which greatly simplifies the analysis required. This assumption has been used in several publications [21, 28] to model an aperture-coupled patch lens antenna.

In order to determine the radiation performance of the lens antenna, the radiation pattern emanated by the patch in the dielectric lens can be calculated from the currents on the patch [31]. This radiation will illuminate the spherical surface of the lens. The far field can be computed based on the equivalent surface electric current density and the equivalent surface magnetic current density on the spherical surface of the lens [21].

When the cost of the lens is a concern, ultra-high-density polyethylene such as Rexolite can be used for the lens material, which is easy to machine and exhibits low loss at millimetre wave frequencies [32]. Applying the design method in Reference [21] to Rexolite material ($\varepsilon_r \approx 2.35$), the length of lens is 64 mm and the radius 50 mm for operation centred at 60 GHz. For substrate A in Figure 6.13 it is preferable for it to have the same permittivity as the lens to minimise the surface wave losses. The design methodology for a proximity-coupled patch in this environment is similar to that when mounted in free space [33, 34]. Thus, for a given set of dielectric materials, the resonant frequency is governed by the length of the patch, and the impedance at resonance is controlled by the offset of the terminated feedline from the centre of the patch. Since a low dielectric constant material was used it was deemed unnecessary to coat the lens with an anti-reflection layer [35].

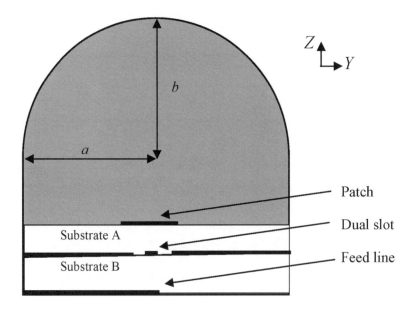

Figure 6.13   Layout for the aperture feed ellipsoidal substrate lens

A general schematic diagram of the single-beam substrate lens antenna is shown in Figure 6.13. The lens is made of Rexolite, a low-cost plastic material.

For generating diffraction-limited patterns, an ellipsoidal lens ($x^2/a^2 + y^2/a^2 + z^2/b^2 = 1$) is chosen, with $a = b\sqrt{(\varepsilon_r - 1)/\varepsilon_r}$, where a and b are the minor and major axes of the ellipsoidal lens, respectively, as shown in Figure 6.13. However, the extension length beyond the major axis is cylindrical instead of elliptical to facilitate the machining process. According to geometrical optics, the length of the cylindrical extension layer should be equal to $b/\sqrt{\varepsilon_r}$ in order to generate parallel rays through the lens when the feed antenna is located on the axis at the far focal point of the ellipsoidal lens. The radiating element used to feed the lens is realised by an aperture-coupled circular polarisation (CP) patch antenna (Figure 6.13).

The feed line of the antenna is built on a high-permittivity substrate B and the patch antenna is printed on a low-permittivity substrate A, which is close to the permittivity of the Rexolite lens. These choices for the substrate are made to increase the bandwidth, as well as to reduce the parasitic radiation losses due to the feed network.

The main advantage of this aperture-coupled patch antenna is that the feeding network and the radiating element are well separated by a ground plane and, thus, the patterns are immune to parasitic radiation [36–38]. Also, the ground plane yields an increased front-to-back (F/B) ratio, which is important since low-permittivity materials are used. Another advantage is that the single line feed structure is well suited for integrated circuit (IC) applications.

If circular polarisation is needed, it can be generated by means of a circular-polarised patch or a cross-shaped slot in the ground plane, which excites two orthogonal modes in a nearly square patch [39]. In particular, the cross-aperture-coupled structure was reported to yield a significant improvement to the CP bandwidth [39]. Research shows that the circular polarisation properties of the structure are robust enough to withstand manufacturing tolerances.

For a silicon lens ($\varepsilon_r = 11.7$) without a matching layer, a typical reflection loss of 1.5 dB is reported in Reference [40], which implies that 30 % of the power is reflected at the lens/air interface. In fact, these reflected rays are not lost, but eventually come out after multiple reflections inside the lens, reducing the directivity and contributing to the final radiation pattern.

6.4 Off-Axis Extended Hemispherical Lens

The dielectric lens also provides mechanical rigidity and thermal stability, and has been used extensively in millimetre and submillimetre wave receivers [41–47]. In Section 6.3, it has been mentioned that if the dielectric lens has the same dielectric constant as the planar antenna wafer, then substrate modes can be eliminated [48]. In addition, antennas placed on dielectric lenses tend to radiate most of their power into the lens side, making the pattern unidirectional on high-dielectric constant lenses. The ratio of powers between the dielectric and air is approximately $\varepsilon_r^{3/2}$ for elementary slot and dipole-type antennas [48], where ε_r is the relative dielectric constant of the lens.

Research works [41, 42, 49] have shown that the directivity of the substrate lens can be controlled by changing the extension length L, as defined in Figure 6.14. In particular, as the extension length increases from the hyperhemispherical length R/n (where R is the radius and n is the index of refraction of the lens), the directivity increases until it reaches a maximum diffraction-limited value.

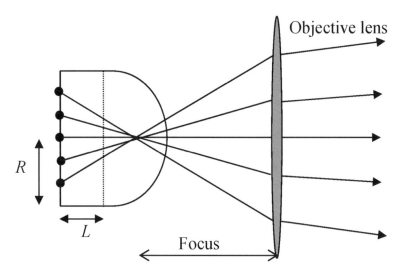

Figure 6.14 A simplified linear imaging array on an extended hemispherical dielectric lens coupled to an objective lens. (Reproduced by permission of © 1997 IEEE [27])

While the directivity increases at higher extension lengths, the pattern-to-pattern coupling value to a fundamental Gaussian beam (Gaussicity) decreases [41]. A Gaussian beam propagating along the z axis as in Figure 6.15, produces a propagating field as:

$$E = \frac{E_0}{z - \mathrm{j}z_0} \exp\left[\frac{-\mathrm{j}kr^2}{2(z - \mathrm{j}z_0)}\right]$$

where $r = (x^2 + y^2)^{1/2}$ and E_0 is a constant.

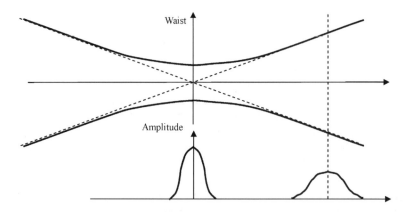

Figure 6.15 Gaussian beam amplitude variations versus distance to the axis follow a Gaussian law, the width of which increases with z while its amplitude decreases with z. (Reproduced by permission of © 1997 IEEE)

Since the double-slot antenna used in Figure 6.13 launches a nearly perfect fundamental Gaussian beam into the dielectric lens, the "Gaussicity" can also be thought of as a measure of the aberrations introduced by the lens.

When extension lengths are up to the hyperhemispherical position, the Gaussicity is nearly 100 % since the hyper-hemi-spherical lens is aplanatic, implying the absence of spherical aberrations, and satisfies the sine condition, which guarantees the absence of a circular coma [50]. As the extension length L increases past R/n, the Gaussicity continuously decreases, which implies the introduction of more and more aberrations. Research results show that for an "intermediate position" between the hyper-hemi-spherical and diffraction-limited extension lengths (e.g. $L/R = 0.32$ to 0.35 for a silicon lens) the Gaussicity decreases by a small amount ($< 10\%$), while the directivity is close to the diffraction-limited value [41, 49]. The choice of an "intermediate position" extension length has resulted in state-of-the-art receivers at 90 and 250 GHz [44, 45, 51].

Figure 6.14 shows the off-axis performance of extended hemispherical dielectric lenses. A ray optics/field-integration formulation in Reference [41] can be used to find the solutions for the radiation patterns and Gaussian coupling efficiencies. Briefly, the radiation of the feed antenna is ray traced to find the fields immediately exterior to the lens surface. For a given ray, the fields are decomposed into TE/TM components at the lens/air interface, and the appropriate transmission formulas are used for each mode. The equivalent electric and magnetic currents are found directly from the fields, and a standard diffraction integral results in the far-field lens patterns [52].

In most applications the dielectric lens will be coupled with a quasi-optical system, and Figure 6.14 shows the dielectric lens coupled to an objective lens. If the Gaussian beams emanating from the dielectric lens are well characterised, then these beams can easily be traced through a quasi-optical system [41] or, for greater accuracy, the patterns emanating from the dielectric lens could be used with electromagnetic (EM) ray-tracing techniques to find the fields across the aperture of the objective lens. Then a Fourier transform will yield the far-field patterns from the objective lens/dielectric lens system.

Any antenna that illuminates the lens surface with a nearly symmetrical, constant phase beam will produce similar results. The black circle in Figure 6.14 represents a radiation element such as the dipole in Figure 6.16 or the dual slot in Figure 6.13. The array radiation can be calculated by assuming a sinusoidal magnetic current distribution on the dipole/slot and by using an array factor in the E-plane direction [53]. The dimensions of the double-slot antenna are of a length $0.28\,\lambda_{air}$ and a spacing of $0.16\,\lambda_{air}$, for a silicon lens with $\varepsilon_r = 11.7$. The dimensions can be scaled to other dielectric materials using the square root of the dielectric constant.

The wavelength of the sinusoidal magnetic current distribution in the slot is approximately the geometric mean wavelength given by $\lambda_m = \lambda_0/\sqrt{\varepsilon_m}$, where $\varepsilon_m = (1 + \varepsilon_r)/2$ [54]. If the double-slot antennas produce a radiation pattern which is 98 % Gaussian, the dielectric lens should also have a similar radiation pattern unless aberrations are introduced by the lens. Note that the patterns radiated to the air side are broader and contain 9.0 % of the total radiated power for a silicon lens. The theoretical technique for analysing the lens radiation patterns is an expanded version of the electromagnetic ray-tracing technique presented in Reference [41].

The lens antenna can be used to launch multiple beams, by printing a multi-element under the base of the lens [37, 40, 55]. As can be seen in Figure 6.17, an array at the back of the lens is used to provide efficient coverage. The scan angle depends on the off-axis displacement X/a, where X is the off-axis distance in Figure 6.17 and a is the minor axis of the designed ellipsoidal lens.

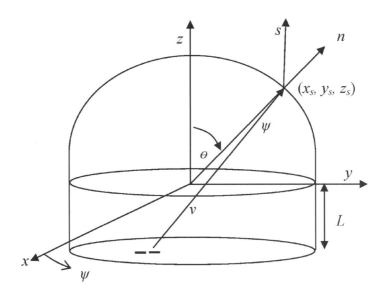

Figure 6.16   The dipole feed lens geometry used for the off-axis theoretical computations

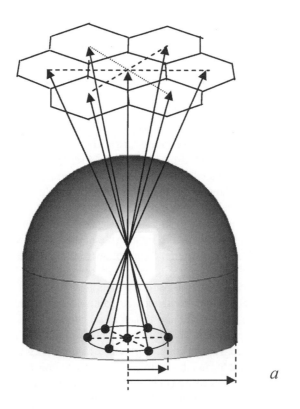

Figure 6.17   Multiple-beam launching through the substrate lens antenna [22]

For wireless communications, one of the most important features for multiple-beam antennas is scan coverage. As demonstrated in Reference [40], the off-axis total internal reflection loss is the limiting factor in the design of larger multiple-beam arrays on substrate lenses. For the present CP design, another possible limitation is off-axis depolarisation.

The peak directivity drops quickly as off-axis displacement increases. In order to launch beams with equal radiation power density and reduce reflection losses, the effect of the extension length L has been numerically investigated [22], and the optimum position has been found to lie around $L \approx a/\sqrt{\varepsilon_r}$. This seems to correspond to the "intermediate" position previously observed for extended hyperhemispherical lenses [40, 56].

6.5 Planar Lens Array

In many applications an RF receiver or transmitter element must be coupled with one or more antennas to focus or distribute RF power. At microwave frequencies, diode and transistor elements have been successfully integrated with photolithographically produced planar antennas. However, at millimetre and submillimetre wavelengths there are three major challenges to many of these lower frequency structures:

1. If the antenna is integrated on a substrate with a dielectric constant greater than 1, unless it is very thin (< 0.1 wavelength), much of the radiated power will flow into modes in the substrate rather than into modes in the air [57].
2. Many planar antennas have low directivity and therefore require very fast optics (low f number) for beam shaping and matching to higher gain systems.
3. Most planar antennas have little or no tuning capability, making matching to the transmit or receive element difficult.

For the moment there does not seem to be an easy way of incorporating millimetre wave adjustable tuning elements into planar antenna structures at submillimetre wavelengths, so it is essential that the antenna itself be well matched to the non-linear device at its terminals over the desired operating band.

The design described in this section combines the features of the dielectric lens antenna in the array configuration and is called the "discrete lens array (DLA)". It has multiple beams with a single spatial feed and can be designed to have dual linear polarization. The lens array is made using standard printed circuit technology and is light weight [58]. As shown in Figure 6.18, a standard N-element antenna array followed by a feed network is replaced by a discrete lens array in which N array element pairs perform a Fourier transform operation on the incoming wave front, and M receivers are placed on a focal surface (when M is smaller than N). The lens array can include integrated amplifiers in each element.

The unit element of the lens array consists of two antennas, interconnected with a delay line. The length of the delay varies across the array, such that an incident plane wave is focused on to a focal point in the near field on the feed side of the array, as in Figure 6.18. Plane waves incident from different directions are focused on to different points on the focal surface, where receiving antennas and circuitry are placed to sample the image, which is a discrete Fourier transform of the incoming wave front. The discrete lens posesses improved focusing properties over some dielectric lenses and reflector antennas, as it can be designed for low sidelobe levels at large pointing angles.

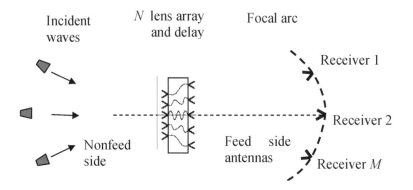

Figure 6.18 The schematic of a lens antenna array. The coupling between the transmission lines on the two sides is accomplished through resonant slots in the common ground plane. The orientation of the patches allows isolation between the two sides of the lens; each frequency is received on the feed side with one polarisation and is radiated from the other side of the lens in the orthogonal polarisation [59]

When multiple receivers correspond to multiple antenna radiation pattern beams, this enables beam-steering and beamforming with no microwave phase shifters. In a multipath environment, each of the reflected waves is focused on to a different receiver, giving angle diversity. Likewise, when transmitters are placed at feed points on the focal surface, multiple beams are radiated, since the lens is linear and superposition of the beams applies. Discrete lenses allow the presence of several simultaneous beams at different angles, with a simpler feed structure than phased arrays.

For the discrete lens, there are two arrays of antennas with transmission lines connecting each radiating element between the two sides. One side is called the *radiating side* and generates the far-field pattern of the lens, while the other side, called the *feed side*, faces the feeds. The transmission lines are of different electrical lengths for each element; the larger delay at the central element with respect to the external ones mimics an optical lens, thicker in the centre and thinner in the periphery. Together with the electrical lengths of the lines, the positions of the array elements on the feed side also determines the focusing properties of the lens. This allows for a design with up to two perfect focal points lying on a focal arc or with a cone of best focus [60]. The two degrees of freedom are in the positions of the elements on the feed side and the electrical lengths of the transmission lines connecting the two sides. The main design constraint is the equality of the path length from the feed to each element on the radiating side of the lens.

The position of the elements on the radiating side dictates the features of the far-field radiation pattern as with a traditional array; the spacing and type of elements are chosen to satisfy the radiation specifications such as grating lobes, sidelobes and beamwidth.

Several feed antennas placed on this focal arc spatially feed the lens, generating a beam in each different direction. Such a feature inherently allows the presence of several independently controlled simultaneous beams. Figure 6.19 shows a schematic of a planar lens with several feeds at different angles with respect to the "optical axis" of the system. The feed positioned on the focal arc at an angle ϕ generates a radiation pattern with the main beam at an angle ϕ, in the far field of the planar lens.

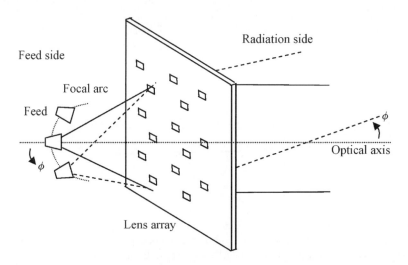

Figure 6.19 Schematic of a planar discrete lens array with several independent feeds on its focal arc. Each feed controls a radiation pattern with the main beam pointing at a different angle off the boresight

The use of a lens array in place of a phased array in multibeam applications presents several advantages arising from the spatial feed concept. The spatial feed allows a multibeam configuration with only minor modifications in the system design, avoiding the high complexity of a feed network.

A phased array in this application would require a multilayer feed structure such as a Butler matrix [61]. Phased array feed networks also have bandwidth limitations due to the phase shifters and their impedance-matching requirements, while the bandwidth of a lens array is limited only by their antenna elements.

Assuming the same power is radiated in the far field for both a phased array and a lens array, the input power requirements for a phased array (PA) with a corporate feed network can be calculated by considering the losses of transmission lines, power dividers and phase shifters, whereas the input power requirements for a discrete planar lens array (LA) can be calculated by considering the losses of transmission lines, path loss and spill-over. Figure 6.20 shows the comparison of these two antennas. More details can be found in Reference [62].

In particular, the main loss in a phased array is due to the power dividers used in the corporate feed network, resulting in dependence on the number of elements, while the main loss in a discrete lens is due to path losses in free space, which increase only negligibly with the lens size. Moreover, a planar lens accomplishes the same functions as a dielectric lens in principle, but presents some advantages. Planar lenses are fabricated using standard PCB technology, making them lightweight, easy to manufacture and easy to optimise for large scan angles [63]. Unlike a dielectric lens, input and output polarisations are a design parameter for planar discrete lenses, allowing different polarisations on the feed and radiating sides of the array.

The planar lens described here is a one degree-of-freedom lens, designed for one perfect focal point on the optical axis. The position of the elements on the feed side are the same as

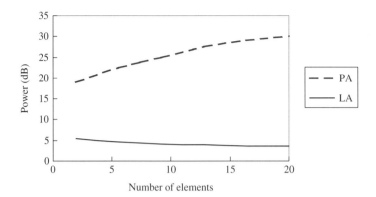

Figure 6.20   Power requirements as a function of the number of elements for phased arrays (PAs) and discrete planar lens arrays (LAs). Together with the total amount of power required to give an equivalent performance, individual contributions to both systems are also illustrated [62]

for the radiating side, leaving the length of the transmission lines connecting the elements as the only design parameter for the focusing properties. The lens has dual-polarisation, dual-frequency patch antenna elements on both sides of a rectangular lattice, with a separation of three-quarters to one free space wavelength between the elements.

A lens array antenna with a multibeam can reduce the multipath fading effect. To measure its performance, the lens is placed in a simple controllable multipath environment consisting of a single metal reflector in an anechoic chamber, as shown in Figure 6.21. The reflector is translated in the x direction over three free space wavelengths. The position of the reflector is tuned so that at $x = 0$, the reflected wave from the transmitting horn falls into the second null of the lens antenna pattern for a receiver on the line-of-sight axis.

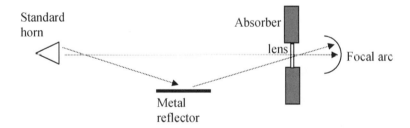

Figure 6.21   Multipath measurement environment [59]

6.6 Metal Plate Lens Antennas

Metal plate lens antennas, also referred to as artificial lens, are attractive and provide some additional gain as dielectric lens. An artifical lens consists of stacked parallel-plate waveguides filled with some low-loss dielectric-like foam. The distance between the plates is chosen to minimise ohmic losses and ensure TE_1 mode operation.

Since the equivalent local refractive index of the waveguides is less than one, each plays a role to increase the phase velocity of the wave propagating through it. The metal plate lens operates by imparting to the waves an increased phase velocity rather than the slower velocity of a dielectric lens described in Sections 6.1 to 6.5. This is due to the fact that electromagnetic waves confined in waveguides, assume a wavelength and phase velocity that are greater than those of free space [64]. This property is acquired by waves confined between parallel conducting plates. The electric field vector is transverse and parallel to the plates. These are spaced apart by a distance e, which should be larger than one half-wavelength and smaller than one wavelength. This condition is required to produce a medium with an refractive index less than unity and the TE_1 single-mode propagation. With such a mode of propagation constrained focusing lens can be produced. This antenna must have a particular profile to transform a spherical wave to a plane wave at the output side.

To design the antenna, the index of refraction of such a lens [65] must first be evaluated:

$$n = \sqrt{1 - \frac{\lambda}{2e\sqrt{\varepsilon_r}}}$$

where e is the spacing between the metal plates and ε_r is the permittivity of the dielectric between the plates (foam in this case). The value of e depends on a choice that gives minimum metal and dielectric losses. Once n is fixed, a plane wave needs to be provided at the output side of the antenna. The plane wave is used in order to avoid phase errors (aberrations) in the radiation pattern. These errors are responsible for high sidelobe levels, thereby producing false detections in the application considered here. The focusing system can be optimised using the geometrical optic (GO). This approach consists of cancelling the phase or the electrical path length between a general ray and the central ray. The range of the design lies between $\pm 15°$ (see Figure 6.22). Consequently, the resultant lens can operate with two feed positions ($\pm 15°$). For the feed placed at $0°$, a better focus can be obtained by varying the distances between the plates. The inner profile of the lens is an ellipse with two foci on $\pm 15°$. The outer profile is also an ellipse shifted by d_0 (central thickness) from the inner one.

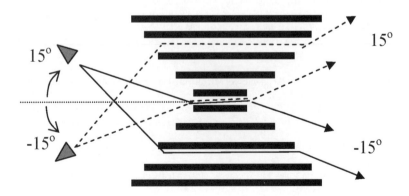

Figure 6.22   Lens made from multiple metal plates [66]

The focal length is governed by geometry and can be obtained from the beamwidth W and desired lens diameter D as:

$$L = 1 - \frac{D}{2 \tan (W_{\text{E-plane}}/2)}$$

The gain can be found as:

$$G = 10 \log_{10}(4.5 A_{e\lambda} A_{h\lambda}) \text{ dB over the dipole}$$

where $A_{e\lambda}$ is the aperture dimension in wavelengths in the E-plane, and $A_{h\lambda}$ is the aperture dimension in wavelengths in the H-plane.

In addition to the geometrical optic method, it is also possible to design a metal plate lens based on Maxwell's equations and co-ordinate transformation. In this case, the lens is considered as a medium with discontinuities described by Dirac delta functions.

The metal lens is smaller than equivalent metal reflecting antennas. Also, these types of lens can focus in both the E- and H-planes. Finally, the lens is adaptable to the requirements of different dimensions and different specifications. Because of its compatibility with planar technology that can be used for a primary source, an artificial lens in foam technology has been found to have good potential for a low-cost solution [67].

References

[1] Germain Chartier, '*Introduction to Optics*', Springer Science and Business Media, Inc., 2005.
[2] R. K. Luneberg, US Patent 2,328,157, 31 August 1943.
[3] J. Thornton, 'Wide-Scanning Multi-layer Hemisphere Lens Antenna for Ka Band', *IEE Proceedings of Microwave Antennas and Propagation*, **153**(6), December 2006, 573–578.
[4] M. Rayner, 'Use of Luneburg Lens for Low Profile Applications', *Datron/Transco Inc. Microwave Product Digest*, December 1999.
[5] R. Donelson, M. O'Shea and J. Kot, 'Materials Development for the Luneburg Lens', International Square Kilometre Array Conference, Geraldton, Australia, 27 July–2 August 2003.
[6] J. Thornton, 'Scanning Ka-Band Vehicular Lens Antennas for Satellite and High Altitude Platform Communications', 11th European Wireless Conference, Nicosia, Cyprus, 10–13 April 2005, pp. 1–2.
[7] B. Schoenlinner, X. Wu, J. P. Ebling, G. V. Eleftheriades and G. M. Rebeiz, 'Wide-Scan Spherical-Lens Antennas for Automotive Radars', *IEEE Transactions on Microwave Theory and Techniques*, **50**(9), 2002, 2166–2175.
[8] Kazuhiro Uehara, Kazuhito Miyashita, Ken-Ichiro Natsume, Kouki Hatakeyama and Koji Mizuno, 'Lens-Coupled Imaging Arrays for the Millimeter- and Submillimeter-Wave Regions', *IEEE Transactions on Microwave Theory and Techniques*, **40**(5), May 1992, 806–811.
[9] G. M. Rebeiz, D. P. Kasilingam, Y. Guo, P. A. Stimson and D. B. Rutledge, 'Monolithic Millimetre-Wave Two-Dimensional Horn Imaging Arrays', *IEEE Transactions on Antennas and Propagation*, **38**, September 1990, 1473–1482.
[10] D. M. Pozar, 'Five Novel Feeding Techniques for Microstrip Antennas', *IEEE Antennas and Propagation Society International Symposium Digest*, June 1987, 920–923.
[11] M. I . Aksun, S . Chuang and Y. T. Lo, 'On Slot-Coupled Microstrip Antennas and Their Applications to CP Operation – Theory and Experiment', *IEEE Transactions on Antennas and Propagation*, **38**, August 1990, 1224–1230.
[12] J. R. James, P. S. Hall and C. Wood, '*Microstrip Antenna Theory and Design*', Peter Peregrinus Ltd, London, 1981, Chapter 4.
[13] M. V. Schneider, 'Microstrip Lines for Microwave Integrated Circuits', *Bell System Technical Journal*, **48**, May–June 1969, 1421–1444.
[14] A. K. Bhattacharyya, 'Characteristics of Space and Surface Waves in a Multilayered Structure', *IEEE Transactions on Antennas and Propagation*, **38**, 1990, 1231–1238.

[15] U. Sangawa, K. Takahashi, T. Urabe, H. Ogura and H. Yabuki, 'A Ka-Band High-Efficiency Dielectric Lens Antenna with a Silicon Micromachined Microstrip Patch Radiator', *IEEE MTT-S International Digest*, **1**, May 2001, 389–392.

[16] M. Al-Tikriti, S. Koch and M. Uno, 'A Compact Broadband Stacked Microstrip Array Antenna Using Eggcup-Type of Lens, *IEEE Microwave and Wireless Components Letters*, **16**(4), April 2006, 230–232.

[17] Wikipedia, http://www.wikipedia.org/

[18] D. F. Filipovic, S. S. Gearhart and G. M. Rebeiz, 'Double-Slot Antennas on Extended Hemispherical and Elliptical Dielectric Lenses', *IEEE Transactions on Microwave Theory and Techniques*, October 1993.

[19] D. F. Filipovic, S. G. Gearhart and G. M. Rebeiz, 'Double-Slot Antennas on Extended Hemispherical and Elliptical Silicon Dielectric Lenses', *IEEE Transactions on Microwave Theory and Techniques*, **41**, October 1993, 1738–1749.

[20] G. V. Eleftheriades, Y. Brand, J.-F. Zurcher and J. R. Mosig, 'ALPSS: A Millimetre-Wave Aperture-Coupled Patch Antenna on a Substrate Lens', *Electronics Letters*, **33**, January 1997, 169–170.

[21] X. Wu, G. V. Eleftheriades and E. Van Deventer, 'Design and Characterization of Single and Multiple Beam mm-Wave Circularly Polarized Lens Antennas for Wireless Communications', IEEE AP-S International Antennas and Propagation Symposium, Orlando, Florida, July 1999, pp. 1200–1204.

[22] X. Wu, G. V. Eleftheriades and T. E. van Deventer-Perkins, 'Design and Characterization of Single- and Multiple-Beam mm-Wave Circularly Polarized Substrate Lens Antennas for Wireless Communications', *IEEE Transactions on Microwave Theory and Techniques*, **49**(3), March 2001, 431–441.

[23] D. F. Filipovic, S. S. Gearhart and G. M. Rebeiz, 'Double-Slot Antennas on Extended Hemispherical and Elliptical Silicon Dielectric Lenses', *IEEE Transactions on Microwave Theory and Techniques*, **41**(10), October 1993, 1738–1749.

[24] M. Qiu and G. V. Eleftheriades, 'Highly Efficient Unidirectional Twin Arc-Slot Antennas on Electrically Thin Substrates', *IEEE Transactions on Antennas and Propagation*, **52**(1), January 2004, 53–58.

[25] D. F. Filipovic and G. M. Rebeiz, 'Double-Slot Antennas on Extended Hemispherical and Elliptical Quartz Dielectric Lenses', *International Journal of Infrared Millimetre Waves*, **14**, 1993, 1905–1924.

[26] J. Zmuidzinas and H. G. LeDuc, 'Quasioptical Slot Antenna SIS Mixers', *IEEE Transactions on Microwave Theory and Techniques*, **40**, September 1992, 1797–1804.

[27] D. F. Filipovic, G. P. Gauthier, S. Raman and G. M. Rebeiz, 'Off-Axis Properties of Silicon and Quartz Dielectric Lens Antennas', *IEEE Transactions on Antennas and Propagation*, **45**, May 1997, 760–766.

[28] P. Otero, G. V. Eleftheriades and J. R. Mosig, 'Integrated Modified Rectangular Loop Slot Antenna on Substrate Lenses for Millimetre- and Submillimeter-Wave Frequencies Mixer Applications', *IEEE Transactions on Antennas and Propagation*, **46**, October 1998, 1489–1497.

[29] D. M. Pozar, 'Microstrip Antennas', *Proceedings of IEEE*, **80**, January 1992, 79–91.

[30] Andrew P. Pavacic, Daniel Llorens del Río, Juan R. Mosig *et al.*, 'Three-Dimensional Ray-Tracing to Model Internal Reflections in Off-Axis Lens Antennas', *IEEE Transactions on Antennas and Propagation*, **54**(2), February 2006, 604–612.

[31] D. M. Pozar, 'Radiation and Scattering from a Microstrip Patch on an Uniaxial Substrate', *IEEE Transactions on Antennas and Propagation*, **AP-35**, June 1987, 613–621.

[32] P. F. Goldsmith, *'Quasi-optical System'*, IEEE Press, Piscataway, New Jersey, 1998, p. 82.

[33] D. M. Pozar and S. M. Voda, 'A Rigorous Analysis of a Microstripline-Fed Patch Antenna', *IEEE Transactions on Antennas and Propagation*, **35**, December 1987, 1343–1349.

[34] D. M. Pozar and B. Kaufman, 'Increasing the Bandwidth of a Microstrip Antenna by Proximity Coupling', *Electronics Letters*, **23**, April 1987, 368–369.

[35] L. Mall and R. B. Waterhouse, 'Millimetre-Wave Proximity-Coupled Microstrip Antenna on an Extended Hemispherical Dielectric Lens', *IEEE Transactions on Antennas and Propagation*, **49**(12), December 2001, 1769–1772.

[36] G. V. Eleftheriades, Y. Brand, J. Zürcher and J. R. Mosig, 'ALPSS: A Millimetre-Wave Aperture-Coupled Patch Antenna on a Substrate Lens', *Electronics Letters*, **33**(3), January 1997, 169–170.

[37] K. Uehara, K. Miyashita, K. I. Natsume, K. Hatakeyama and K. Mizuno, 'Lens-Coupled Imaging Arrays for the mm and Sub-mm-Wave Regions', *IEEE Transactions on Microwave Theory and Techniques*, **40**, May 1992, 806–811.

[38] X. Wu, G. V. Eleftheriades and E. van Deventer, 'A mm-Wave Circularly Polarized Substrate Lens Antenna for Wireless Communications', Symposium on *'Antenna Technology and Applied Electromagnetics (ANTEM)'*, Ottawa, Canada, August 1998, pp. 595–598.

[39] T. Vlasits, E. Korolkiewicz, A. Sambell and B. Robinson, 'Performance of a Cross-Aperture Coupled Single Feed Circularly Polarized Patch Antenna', *Electronics Letters*, **32**(7), March 1996, 612–613.

[40] D. F. Filipovic, G. P. Gauthier, S. Raman and G. M. Rebeiz, 'Off-Axis Properties of Silicon and Quartz Dielectric Lens Antennas', *IEEE Transactions on Antennas and Propagation*, **45**, May 1997, 760–766.

[41] D. F. Filipovic, S. S. Gearhart, and G. M. Rebeiz, 'Double Slot Antennas on Extended Hemispherical and Elliptical Silicon Dielectric Lenses', *IEEE Transactions on Microwave Theory and Techniques*, **41**, October 1991, 1738–1749.

[42] T. H. Buttgenbach, 'An Improved Solution for Integrated Array Optics in Quasioptical Millimeter and Submillimeter Waves Receivers: The Hybrid Antenna', *IEEE Transactions on Microwave Theory and Techniques*, **41**, October 1991, 1750–1761.

[43] J. Zmuidzinas, 'Quasioptical Slot Antenna SIS Mixers', *IEEE Transactions on Microwave Theory and Techniques*, **40**, September 1991, 1797–1804.

[44] G. P. Gauthier, W. Y. Ali-Ahmad, T. P. Budka, D. F. Filipovic and G. M. Rebeiz, 'A Uniplanar 90 GHz Schottky-Diode Millimetre Wave Receiver', *IEEE Transactions on Microwave Theory and Techniques*, **43**, July 1995, 1669–1672.

[45] S. S. Gearhart and G. M. Rebeiz, 'A Monolithic 250 GHz Schottky Diode Receiver', *IEEE Transactions on Microwave Theory and Techniques*, **42**, December 1994, 2504–2511.

[46] H. Z. Zirath, C.-Y Chi, N. Rorsman and G. M. Rebeiz, 'A 40-GHz Integrated Quasi-optical Slot HFET Mixer', *IEEE Transactions on Microwave Theory and Techniques*, **42**, December 1994, 2492–2497.

[47] A. Skalare, H. van de Stadt, Th. de Graauw, R. A. Panhuyzen and M. M. T. M. Dierichs, 'Double-Dipole Antenna SIS Receivers at 100 and 400 GHz', Proceedings of 3rd International Conference on *'Space Terahertz Technology'*, Ann Arbor, Michigan, March 1992, pp. 222–233.

[48] D. B. Rutledge, D. P. Neikirk and D. P. Kasilingam, 'Integrated Circuit Antennas', in *'Infrared and Millimetre-Waves'*, Ed. K. J. Button, Academic Press, New York, 1983, Vol. 10, pp. 1–90.

[49] D. F. Filipovic and G. M. Rebeiz, 'Double Slot Antennas on Extended Hemispherical and Elliptical Quartz Dielectric Lenses', *International Journal of Infrared Millimetre Waves*, **14**, October 1991, 1905–1924.

[50] M. Born and E. Wolf, *'Principles of Optics'*, Pergamon, New York, 1959, pp. 252.

[51] D. Filipovic, G. V. Eleftheriades and G. M. Rebeiz, 'Off-Axis Imaging Properties of Substrate Lens Antennas', 5th International Space Terahertz Technology Symposium, Ann Arbor, Michigan, February 1994, pp. 778–787.

[52] C. A. Balanis, *'Antenna Theory: Analysis and Design'*, John Wiley & Sons, Inc., New York, 1982, Chapter 11.

[53] R. S. Elliott, *'Antenna Theory and Design'*, Prentice-Hall, Englewood Cliffs, New Jersey, 1981, Chapter 4.

[54] M. Kominami, D. M. Pozar and D. H. Schaubert, 'Dipole and Slot Elements and Arrays on Semi-infinite Substrates', *IEEE Transactions onAntennas and Propagation*, **AP-33**, June 1985, 600–607.

[55] X. Wu, G. V. Eleftheriades and E. van Deventer, 'Design and Characterization of Single and Multiple Beam mm-Wave Circularly Polarized Substrate Lens Antennas for Wireless Communications', Proceedings of IEEE International Antennas and Propagation Symposium, Orlando, Florida, July 1999, pp. 2408–2411.

[56] D. F. Filipovic, S. S. Gearhart and G. M. Rebeiz, 'Double-Slot Antennas on Extended Hemispherical and Elliptical Silicon Dielectric Lenses', *IEEE Transactions on Microwave Theory and Techniques*, **41**, October 1993, 1738–1749.

[57] D. B. Rutledge, D. P. Neikirk and D. P. Kasilingam, 'Integrated-Circuit Antennas', in *'Infrared and Millimetre Waves'*, Academic Press, New York, 1983, Vol. 10, pp. 1–90.

[58] D. T. McGrath, 'Planar Three-Dimensional Constrained Lenses', *IEEE Transactions on Antennas and Propagation*, **AP-34**, January 1986, 46–50.

[59] Darko Popovic and Stefania Romisch, 'Multibeam Planar Lens Antenna Arrays', *2003 GOMAC Digest*, Tampa, April 2003.

[60] D. T. McGrath, 'Planar Three Dimensional Constrained Lenses', *IEEE Transactions on Antennas and Propagation*, **AP-34**(1), 1986, 46–50.

[61] R. C. Hansen, *'Phased Array Antennas'*, Wiley Series in Microwave and Optical Engineering, Series Ed.,Key Shang, John Wiley & Sons, Inc. New York, 1998, Chapter 10.

[62] S. Römisch, N. Shino, D. Popovic, P. Bell and Z. Popovic, 'Multibeam Planar Discrete Millimeter-Wave Lens for Fixed-Formation Satellites', *IEEE-MTT Microwave Symposium Digest*, **3**, 2003, 1669–1672.

[63] D. Popovic and Z. Popovic, 'Multibeam Antennas with Polarization and Angle Diversity', *IEEE Transactions on Antennas and Propagation*, Special Issue on Wireless Communications, **50**, May 2002, 651–657.

[64] W. E. Kock, 'Metal-Lens Antennas', *Proceedings of the IRE*, **34**, November 1946, 828–836.

[65] J. Ruze, 'Wide-Angle, Metal-Plate Optics', *Proceedings of the IRE*, January 1950, 53–59.

[66] E. Jehamy, G. Landrac, S. Pinel, B. Della, F. Gallee and M. Ney, 'A Compact Constrained Metal Plate Lens for Anti-collision Radar at 76 GHz', International Conference on Antennas, Berlin, 2003.

[67] G. Granet, I. Fenniche, K. Edee, J. P. Plumey, E. Jehamy and M. Ney, 'New Method for Analysis of Constrained Metal Plate', International Conference on *'Electromagnetics for Advanced Applications'*, Italy, 2007.

7

Multiple Antennas

Multiple antennas play a key role at millimetre waves as they can either increase antenna gain or improve antenna diversity. Additionally, multiple antennas can be operated either to maximise the channel throughput (capacity) or to maintain the robustness of the link. At lower frequencies, multiple antennas operating as an array can be used to construct one composite beam so that the elements are in fact electromagnetically coupled to form a single wavefront. By this means a higher gain and directivity can be achieved. By decoupling the antenna elements, multiple beams can be formed so that the array acts to support many different beams or look directions. In this case the beams can be configured either to support multiple communications channels (multiple-input multiple-output, or MIMO), to increase the capacity or to support multiple copies of the same information stream to enhance the robustness of the link (diversity). Coding schemes can also be applied to these arrays and such research [1] is a major area for future communications development [2]. The way in which the antenna is driven at the logical layer is not the primary concern in this text, but instead concentrates on the electromagnetic aspects of the devices. The reader is referred to texts that deal with the signal processing aspects of multiple antenna systems [3]. This section therefore first introduces 60 GHz multibeam antennas, which can be used to increase antenna diversity (link robustness/multiple channels), especially in an indoor environment. Next, the antenna array design is discussed. Then, three types of millimetre wave array are described: the printed array, the waveguide array and the leaky-wave array. Finally, mutual coupling between antenna elements is analysed for design consideration purposes.

7.1 The 60 GHz Multibeam Antenna

In a multipath environment, antennas with a narrow beam can be used to reduce the number of multipath signals and therefore to minimise the root-mean-square delay spreads. Moreover, a narrow-beam antenna has a high directivity that directs or confines the power or reception in a given direction and thus extends the communication range. Furthermore, the gain of the antennas can partly reduce the required gain of millimetre wave power amplifiers, by supplying more captured power to the output terminals of the antenna, so the power consumption in millimetre wave circuitry will potentially be reduced.

Millimetre Wave Antennas for Gigabit Wireless Communications   Kao-Cheng Huang and David J. Edwards
© 2008 John Wiley & Sons, Ltd

Conventional millimetre wave links can be classified based on whether or not an uninterrupted line-of-sight (LOS) is established between the transmitter and receiver. For indoor applications, non-line-of-sight (NLOS) scenarios, also called "diffuse links", are very common [5]. Conventional millimetre wave communication systems, whether LOS or NLOS, mostly employ a single antenna. This section describes a way to improve the performance by using a multibeam directional array, which utilises multiple elements that are pointed in different directions [5]. Such an angle-diversity antenna array can have an overall high directivity and large information capacity.

The concept also offers the possibility of reducing the effects of co-channel interference and multipath distortion, because the unwanted signals are angularly filtered out (by the individual narrow beams). The multibeam antenna array can be implemented using multiple dielectric rods that are oriented in different directions. A conventional rod antenna is fed by waveguides [6], an arrangement that is too bulky for the particular array structure under consideration here. The more attractive approach for the dielectric rod configuration uses a patch-fed method [7] to make a rod array, which has the advantage that it can be integrated with a variety of planar circuits.

The geometry of the antenna array is shown in Figure 7.1. The central rod antenna (1) is in an upright direction, which is perpendicular to the plane of the patches, while the other rods (2, 3, 4, 5, 6 and 7) are tilted with a polar angle (θ) of 40° relative to the surface normal, towards the plane of the patches. The tilted antennas are rotated with respect to the central rod and have an azimuthal angular spacing ($\Delta\Phi$) of 60°. Seven rod antennas are mounted on a metal plate with corresponding feeds. Each rod has a different angular radiation coverage. When the antennas cover a given spatial area, each rod covers a nominally non-overlapping cell in a similar arrangement to a cellular system, as shown in Figure 7.1 (c).

The centre frequency of the antenna described here is designed for 60 GHz. As shown in Figure 7.2, the rods are made of Teflon® with a 3 mm diameter cylindrical base. The upper part of the rod is tapered linearly to a terminal aperture of a 0.6 mm diameter to reduce minor lobes in the radiation pattern. The total height of the central rod antenna is 20 mm to ensure a high antenna gain. The diameter of the central rod (1) antenna is designed to be 3 mm. Each rod antenna is designed to have a 40° half-power beamwidth. The whole radiation of seven-rod antennas therefore covers a polar angle of approximately 60° with respect to the z axis, which is the axis perpendicular to the plane of the patches.

The fields at the rod surfaces are derived using equivalent electric and magnetic current sheets, and the radiation field is simulated from these currents. The relative electric field pattern E as a function of the polar angle θ from the axis is derived by the following formula [7]:

$$E(\theta) = (\sin \Phi)/\Phi \tag{7.1}$$

where $\Phi = H_\lambda \pi(\cos \theta - 1) - 0.5\pi$ and H_λ is the height of the rod in free space wavelengths.

The tapered rod can be treated as an impedance transformer, and reduces the reflection that would be caused by an abrupt discontinuity [8]. The rods are fed by patches on low-temperature co-fired ceramic (LTCC) substrates (see Chapter 9) [9, 10]. The patch-fed method can adapt rod antennas to most planar circuits. It firstly saves feeding space, and also brings more design flexibility to the array structure. Additionally, it also increases the directivity and bandwidth of conventional patch antennas.

The patches are either circular or elliptical in shape to match the base of the rods. The patches are individually energised by probes connected to coaxial connectors. The probe feed

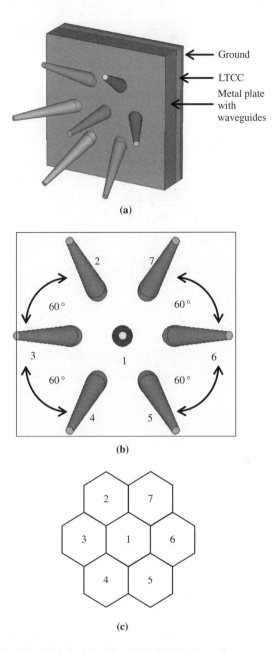

(a)

(b)

(c)

Figure 7.1   (a) Geometry and (b) top view of a seven-rod antenna. (c) Radiation coverage in a hexagonal configuration. (Reproduced by permission of © 2006 IEEE [4])

can provide smaller sidelobes in comparison with microstripline feeds, because the coupling between the antennas and the feeding lines is limited by the ground plane.

The performance of the above antennas was simulated using CST Microwave Studio, which is a simulator based on the finite integration time-domain method. The spacing between

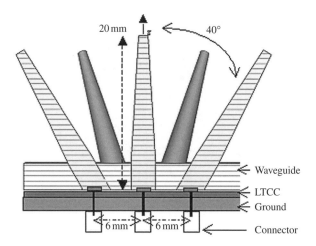

Figure 7.2   Side view of the rod antennas. (Reproduced by permission of © 2006 IEEE [4])

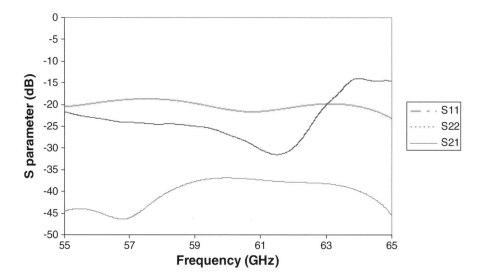

Figure 7.3   Measured S parameters for the upright rod,1 (S11), one of the tilted rods 2 (S22) and their mutual coupling (S21). (Reproduced by permission of © 2006 IEEE [4])

adjacent rods is set to 6 mm to allow enough space between test connectors. As can be seen in Figure 7.3, the antenna can be operated from 55 to 65 GHz with a return loss of about −23 dB. Rod antennas show a broad impedance bandwidth and are suitable for wireless personal area networks such as IEEE 802.15.3c related applications [11]. The couplings between adjacent rods (e.g. 1 and 2; 3 and 2) were measured to be approximately −40 dB. Because of the symmetric configuration, any two pairs of rods with the same relative spacing have similar coupling coefficients. Appropriate grounding of the conducting plate is important to reduce coupling due to surface-wave propagation.

The 60 GHz radiation pattern is shown in a Cartesian plot in Figure 7.4. It depicts the 60 GHz radiation pattern of the upright rod 1 and a tilted rod 3 at $\theta = 0°$ to the plane. The main beam is in the direction of $\theta = 0°$ and $-40°$ for the upright rod and the tilted rod, respectively. It is shown that the maximum gain of rod 3 is radiated in the same direction as the physical axis of the rod. The upright rod has a half-power beamwidth at $\theta = -20°$, while the tilted rod has a half-power beamwidth between $\theta = -14°$ and $-53°$. Because of the asymmetric shape of the tilted rod, one side has a longer height than the other. The side with the long height can radiate and receive more energy than the short side. Thus, its radiation pattern is asymmetric, as shown in Figure 7.4. The curves are normalised to the pattern of the upright rod to show the relative power. For the tilted rod, there is a sidelobe at 13° with a level of about -8.5 dB. Measurement results for the proposed antennas were obtained to confirm the theoretical predictions.

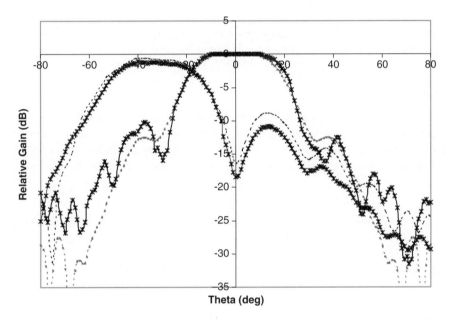

Figure 7.4 Simulated (--) and measured (-*-) radiation patterns at phi $= 0°$ plane as a function of the elevation angle at 60 GHz when the main beam is in the direction of theta $= 0°$ and $-40°$, respectively. These curves are normalised to that of the central rod 1. (Reproduced by permission of © 2006 IEEE [4])

The frequency response was measured by a comparison method with a V-band standard horn antenna. As can be seen in Figure 7.5, it is observed that average antenna gains of about 11.5 and 9.8 dBi were measured at between 57 and 65 GHz for the upright rod and the tilted rods, respectively. The 3 dB bandwidth of both the upright antenna and the tilted antenna are approximately 19 % of the centre frequency, which is higher than the 11 % bandwidth reported in Reference [12]. By comparing the measured antenna gain and the directivity, the radiation efficiency of the implemented prototype is estimated to be 80 and 73 % for the upright rod and the tilted rods, respectively, while the aperture efficiency is 74 and 57 %, respectively. Note that the main beam, to a good approximation, circularly symmetric in its half-power

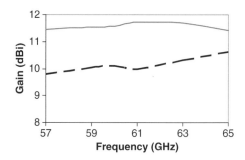

Figure 7.5 Maximum gain of the upright rod (solid line) and tilted rod (dashed line) between 57 and 65 GHz. (Reproduced by permission of © 2006 IEEE [4])

beamwidth region, and therefore the inter-element angle spacing ensures a 3 dB radiation pattern overlapping in any principal cut in the azimuthal plane.

7.2 Antenna Arrays

Antenna arrays are used to direct radiated power towards a desired target. The number, geometrical arrangement and relative amplitudes and phases of the array elements determine the angular pattern that is synthesised. In general, the pattern of a one-dimensional array with an n-element is represented by the complex equation:

$$R = \Sigma \omega_n \exp(jkd_n)$$

where ω_n is the complex excitation, $k = 2\pi/\lambda$ and d_n is the element spacing; the inner product shows the angular variation between k and d_n. By varying this phase shift, the beam position can be scanned.

There are three basic methods that are commonly used to design a high gain, sharp beam array:

1. The uniform array. This means that all the (closely spaced) elements have the same amplitude and phase weight (excitation). Such a configuration has a very narrow mainlobe as well as a very high sidelobe. For an eight-element array, the amplitude weight can be written as:

$$W_{N=8} = [1, 1, 1, 1, 1, 1, 1, 1]$$

2. The binomial array [13]. The weights of an N-element array are the binomial coefficients:

$$\frac{(N-1)!}{n!(N-n-1)!}, \quad n = 0, 1, 2, \ldots, N{-}1$$

For an eight-element array the amplitude weight can be written as:

$$W_{N=8} = [1, 7, 21, 35, 35, 21, 7, 1]$$

Such a configuration has a wide mainlobe but very low sidelobes in its radiation pattern.

3. The Dolph–Chebyshev array [13]. The weights of an N-element array are defined by the Chebyshev polynomial of degree $N - 1$ [13]:

$$
w_N(\psi) = T_{N-1}(x) = \begin{cases} (-1)^{N-1} \cosh[(N-1)\arccos h|x|], & x < -1 \\ \cos[(N-1)\arccos x] & |x| \le 1 \\ \cosh[(N-1)\arccos hx] & x > 1 \end{cases}
$$

This configuration achieves a compromise between mainlobe width and sidelobe attenuation. For a given minimum sidelobe level, the narrowest possible mainlobe width can be designed using this Dolph–Chebyshev array.

This section will mainly focus on the uniform array, which is simple to implement at millimetre waves. The radiation pattern of such an array can be treated as a multiplication of the radiation pattern of a single element with an array factor. This can be expressed as:

$$
R_{\text{array}} = A R_{\text{single}}
$$

where R_{array} is the radiation function of an array, R_{single} is the radiation function of a single element and A is the array factor. An array of identical antennas can modify the single-antenna radiation function by this array factor, which incorporates all the translational phase shifts and relative weighting coefficients of the array elements.

For a uniformly spaced one-dimensional array along the x axis (see Figure 7.6), the array factor can be written as:

$$
A(\theta, \phi) = \sum_n a_n e^{jknd \sin \theta \cos \phi} \tag{7.2}
$$

where n is the number of elements, a_n is the amplitude of the nth element, k is the wave vector and d is the distance between the elements. For the x axis array, the azimuthal angle varies over $-\pi \le \phi \le \pi$, but the array response is symmetric in ϕ and can be evaluated only for $0 \le \phi \le \pi$. By defining a variable *digital wavenumber* $\psi = d \sin \theta \cos \phi$, the array factor (7.2) can be rewritten as:

$$
A(\psi) = \sum_n a_n e^{jn\psi}
$$

$A(\psi)$ is periodic in ψ with period 2π, and therefore it is sufficient to know it to within one Nyquist interval, i.e. $-\pi \le \psi \le \pi$.

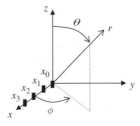

Figure 7.6   A typical array configuration along the x axis

However, in the xy plane, as the azimuthal angle ϕ varies from 0 to π, the term ψ varies from kd to $-kd$. Thus the overall range of variation of ψ is $-kd \leq \Psi \leq kd$. This range is also called the "visible region". Depending on the distance between elements d, the visible region can be larger than, equal to or smaller than one Nyquist interval:

$$d = \lambda/2 \quad \rightarrow \quad kd = \pi \quad \rightarrow \quad \psi_{\text{vis}} = 2\pi \text{ (equal to Nyquist)}$$

$$d < \lambda/2 \quad \rightarrow \quad kd < \pi \quad \rightarrow \quad \psi_{\text{vis}} < 2\pi \text{ (smaller than Nyquist)}$$

$$d > \lambda/2 \quad \rightarrow \quad kd > \pi \quad \rightarrow \quad \psi_{\text{vis}} > 2\pi \text{ (larger than Nyquist)}$$

In the case when the spacing is larger than the Nyquist interval, the further problem of *grating lobes* arises. In this case, at certain angles the contribution from the elements destructively interferes and nulls are produced. The resulting pattern can have a series of these nulls, and the pattern is said to possess grating lobes similar to the case of diffraction gratings in optics.

For a one-dimensional uniform array, the array factor will be of the form:

$$A(\psi) = \frac{\sin(N\psi/2)}{N\sin(\psi/2)} e^{j(N-1)\psi/2}$$

In an example of an eight-element array as shown in Figure 7.7, the radiation pattern can be varied as a function of the distance d. Three cases will be used to demonstrate differences.

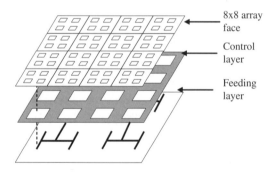

Figure 7.7   Three-tile construction of the patch array

First, when $d = \lambda/2$ (e.g. 2.5 mm for 60 GHz), the array pattern can be drawn as in Figure 7.8 (a). Then, when d is reduced to $\lambda/4$, the beamwidth of the mainlobe increases, and the number and magnitude of the sidelobes are reduced (see Figure 7.8 (b)). Finally, when d is increased to λ, the beamwidth of the mainlobe reduces but grating lobes appear on both sides of the mainbeam, due to the destructive interference effect described previously (see Figure 7.8 (c)).

When there are more than six elements ($N > 6$) in an array, the half-power beamwidth of such an array at broadside is approximately:

$$\Delta\phi_{3\text{dB}} = 0.886 \frac{\lambda}{Nd}$$

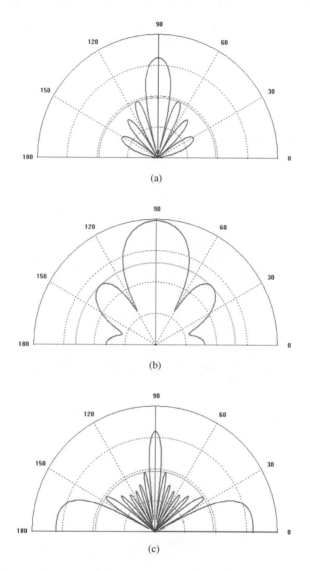

Figure 7.8 Radiation pattern for an uniform array with different d: (a) $d = \lambda/2$, (b) $d = \lambda/4$ and (c) $d = \lambda$

A convenient parameter to use is the sidelobe level, which is the ratio of the power density in the sidelobe to the power density in the mainlobe. The sidelobe level varies as the excitation and the number of element changes or beam direction changes. Figure 7.9 shows an 8×8 array with the same phase and amplitude in each element, and its radiation pattern at 60 GHz. The main beam is aligned in the broadside direction.

Figure 7.10 gives the mainlobe level and sidelobe level versus the number of elements for a patch array on a ceramic substrate ($\varepsilon_r = 10$), for the element distances 2 and 2.5 mm at 60 GHz.

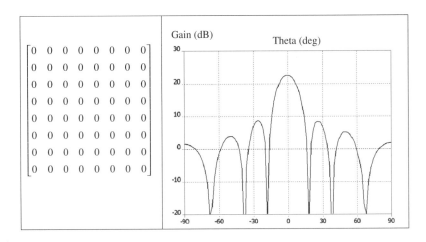

Figure 7.9 Phase arrangement for an 8×8 array with the same amplitude for all elements and its E-plane radiation pattern (Gain in dBi)

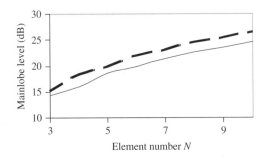

Figure 7.10 Mainlobe level versus element number N for an $N \times N$ patch array on a ceramic substrate with a separation distance $d = 2\,\text{mm}$ (solid line) and 2.5 mm (slot line) at 60 GHz (Gain in dBi)

Arrays of up to 10×10 elements have been found to exhibit sidelobe ratio degradation. As the number of elements increases, the mainlobe level increases.

By Changing the phase arrangement of an array in the x axis (e.g. Figure 7.11), the beam can be tilted. Because of the asymmetric beam pattern with respect to the ground plane of the patch array, one of the first sidelobe levels increases while the other first sidelobe level on the other side of the main beam decreases. Overall, the number of sidelobes remains constant when the beam is steered away from broadside.

7.3 Millimetre Wave Arrays

7.3.1 Printed Arrays

Printed circuit antennas are simple in structure and easy to fabricate by lithography. They are low-profile, lightweight, low-cost devices that are well suited to be used as radiating elements for planar and conformal array antennas. In active arrays they allow convenient

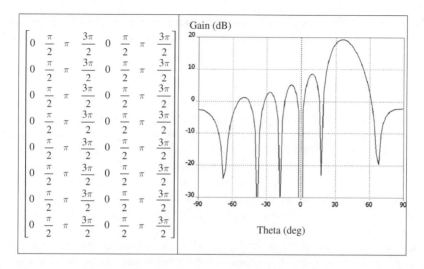

Figure 7.11 Phase arrangement for an 8×8 array with the same amplitude in the x direction elements and its radiation pattern in the x direction (Gain in dBi)

integration with active and passive circuits for beam control and signal processing. The best-known printed circuit antennas are microstrip patch resonator antennas, and microstrip dipole antennas. These antennas have been studied extensively during the past ten or twenty years, at microwave frequencies, and can be regarded as well understood. The main problem with microstrip antennas, and in particular simple patch resonator antennas, is that they have a narrow bandwidth, which does not allow for their use in broadband systems. For a general review on microstrip antennas, see Reference [14]. This also discusses broadband techniques and the important question of the design of appropriate feed systems for phased arrays of these antennas, including their integrated versions. The present discussion addresses the millimetre wave aspects of printed arrays, and is concerned primarily with passive microstrip antennas and arrays. Active versions are discussed in Chapter 8 which deals with integrated antennas.

The extension of microstrip antennas into the millimetre wave region is not simply a matter of straightforward wavelength scaling. New problems, as well as new opportunities appear. There are in particular two problems: fabrication tolerances and feed line losses. Both problems are associated primarily with the feed systems of millimetre wave microstrip arrays rather than with the radiating elements themselves. For ease of fabrication, the feed system, in the microstrip, is usually printed with the radiating elements on the same substrate surface [15], or it may be embedded in the substrate [16]. Typically the feed lines have a width in the order of a few tenths of a millimetre for an array operating at a frequency in the 30–100 GHz band, and the required tolerances for these are tighter by an order of magnitude. Thus highly precise fabrication techniques are needed [15]. Furthermore, microstriplines are not low-loss lines, and in large arrays, which are needed to obtain a high-directivity/gain, feed line losses can be substantial, particularly when a complex feed system is used.

Various methods have been studied in recent years to resolve these problems, and in particular the efficiency problem [15, 16]. The results obtained lead to the conclusion that microstrip patch resonator arrays will be useful antennas for the lower millimetre wave band up to a frequency of 100 GHz and possibly 140 GHz [17]. For this band it should be possible to design microstrip antenna arrays with rather high gain that provide good pattern quality, low reflection losses over a bandwidth of several per cent and with acceptable efficiency [17].

An advantage of microstrip arrays at millimetre waves is that their bandwidth limitation can be overcome by the use of electrically thick substrates. In the microwave region, the substrate thickness is typically only a small fraction of a wavelength. An antenna printed on a thin, grounded substrate will have to operate in close proximity to the ground plane, which implies the use of resonator based antennas of high Q in order to raise the radiation resistance to reasonable values. This, in turn, leads to a narrow bandwidth. But, a physically thin substrate can still have an electrical thickness of $\lambda/4$ or more in the millimetre wave region. Hence, there is no need for the use of high-Q resonator antennas, but other antenna configurations that provide much broader bandwidths can be used, for example, printed dipole antennas of large width. The use of thick substrates has the additional advantage that fabrication tolerances become less critical.

As high directivity is generally desired for a millimetre wave antenna, this implies that the use of arrays of microstrip dipole antennas on thick substrates can be used. The design of such arrays involves new problems. Surface waves trapped in the substrate could increase mutual coupling between the array elements. A low-loss feed system is needed that is easy to fabricate but keeps the feed line radiation to a minimum. In the case of a scanned array, the occurrence of blind spots may severely limit the useful scan range. These problems are not easy to solve. An attractive approach to the feed system problem is the use of a two-layer substrate, as investigated by Katehi and Alexopoulos [18].

In Figure 7.12, the strip feed lines are printed on the lower layer, which is electrically thin so that the energy guided by the feed lines is tightly bound and little feed system radiation will be generated. The strip dipoles are printed on the upper layer and are excited by near-field coupling without the need for conductive contact to the feed lines. In the case of a thick substrate, as assumed here, it may be necessary to enhance this coupling with the help of parasitic dipoles [19] embedded in the upper substrate layer or printed with the radiating dipoles on the top substrate surface. The use of such dipoles has the added benefit that it will increase the bandwidth of the antenna.

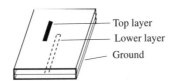

Top layer
Lower layer
Ground

Figure 7.12   Printed strip dipole [18]

7.3.2 Waveguide Arrays

In practice, the architecture of a waveguide array allows the array designer to vary all of the electromagnetic properties of the elements, in order to control the aperture distribution [20–22]. Architectural choices begin at the aperture and dictate how the elements are to be grouped and fed. Behind the aperture can be placed phase and amplitude control, which is followed by a network that combines the signal from the various elements and provides amplitude weighting, time/phase delay and perhaps adaptive control for real-time steering or interference suppression. The control aspect begins with the millimetre wave phase shifters that have been the mainstay of electronic scanning systems since the first arrays. However, recent demands for wideband performance and highly flexible array control, including adaptive and reconfigurable arrays, have highlighted the special features offered by optical and digital control.

An array *brick* can be a single module where the construction process is reduced to assembling the array face one element/module at a time. This has been the established practice for most radar arrays at frequencies up to 75 GHz because of the element size and separation. In this case, the array element modules, which consist of an element and a phase shifter, are inserted into a *manifold* that provides RF power and phase shifter control. The modules can also include active devices, amplifiers and switches, and so may be complete transmit–receive front ends in their own right. In this way, the transmitter and receiver chain is a part of the array face and can offer performance benefits in terms of noise and losses. The RF power division is accomplished in the manifold. This assembly technique is efficient and relatively easy to maintain [20].

With current technology, waveguide array architecture may not be practical nor have the lowest cost at millimetre wave frequencies, and so could be replaced by a brick construction with a number of elements in each brick, or by a tile construction. The reason as to why frequency enters into this selection is that a semi-conductor or superconductor substrate has limited dimensions. As the frequency goes up, it becomes easier to place more devices and elements on the same chip. At these frequencies, the use of a multiple-element "brick and tile" construction becomes a practical proposition.

With high-precision fabrication, waveguide slot arrays (Figure 7.13) can provide excellent pattern control at millimetre wave frequencies. Rama Rao [23] used photolithographic technology to build waveguide longitudinal shunt slot and inclined series slot arrays at 94 GHz.

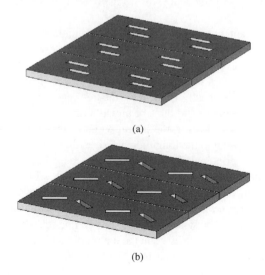

(a)

(b)

Figure 7.13 Waveguide slot array geometries: (a) displaced longitudinal slot array and (b) inclined series slot array [25]

The longitudinal shunt slot array was designed according to the formulas of Yee [24], while the inclined series slot array was based on the analysis of Oliner [25].

Recently research has been undertaken with a view to combining solid-state circuit devices with antennas, primarily Vee strip dipoles and TEM horns [26]. At millimetre wavelengths the power source or pre-amplifier/mixer can be embedded on the same substrate as the antenna. An option on the printed flat TEM horn is the square linear taper horn with dipole feed at the mouth opening; the horn walls are then etched and metallised on to a thick substrate, as in the feed [27]. Modern multilayer substrates such as LTCC and liquid crystal polymer (LCP) can have the waveguide integrated with the substrate, and therefore the waveguide array becomes feasible.

7.3.3 Leaky-Wave Arrays

Most millimetre microstrip and dielectric waveguides have open guiding structures, and energy leakage will occur when the uniformity of these guides is perturbed or they are not excited in the appropriate mode. Leaky waves are fast waves while surface waves are slow waves. This leakage effect may be used to advantage for the design of antennas, by intentionally introducing perturbations in these guides so that they radiate in a controlled fashion. The attractive point to using leaky-wave antennas at millimetre wave frequencies is that they become physically short while still providing a high gain. Another feature of these antennas is that they employ a single series feed, which results in reduced spurious radiation and conductor losses when compared to arrays fed by lossy and bulky corporate-fed networks. This is desirable at millimetre wave frequencies where conductor losses increase dramatically. At the same time, a wide VSWR bandwidth and frequency scanning capabilities can be achieved compared to series-fed resonant antennas (which are essentially narrowband devices). However, the integration of shunt elements for beam steering at a fixed frequency becomes a challenge for printed technology at millimetre wave frequency, due to the need for via holes through the substrate [28, 29].

A general leaky-wave antenna can be made by placing dielectric or conducting strips periodically along a dielectric waveguide [30]. These strips form a grating that perturbs the energy travelling along the guide, exciting leaky modes above the surface that determine the nature of the far-field pattern, which has a sharp beam in the array-effect direction. The axial lengths of the leaky-wave antennas are typically 10 to 50 λ_0 (the free space wavelength); i.e. the antennas are long in the forward direction, and their beamwidth, in the principal plane parallel to the longitudinal axis, is narrow. However, the lateral width of these antennas is small and their beamwidth in the plane normal to the longitudinal axis is therefore wide. Pencil beams with a narrow beamwidth in both principal planes may be achieved by the use of an array with several of these line source antennas in a parallel arrangement (Figure 7.14).

Figure 7.14 (a) and (b) are cross-section views of two such arrays, consisting of antennas that leak due to their asymmetry [31]. It is assumed here that each of the line sources is fed from one end, with an arrangement by which a phase shift can be introduced between successive line sources. The array can then be scanned in the longitudinal plane by varying the frequency, and in the cross plane by varying the phasing. The vertical metal baffles [31] separating the element antennas serve to eliminate blind spots, a major problem that usually arises in arrays of this type, and they also allow the designer to place the line sources close to each other with minimal mutual interaction.

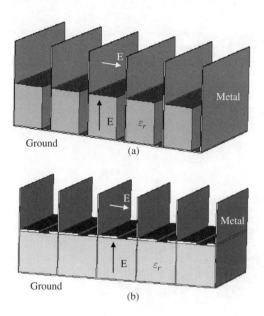

Figure 7.14 Arrays of leaky-wave line sources that can be scanned in one principal plane by frequency variation, and in the other by phasing. The vertical metal baffles permit these arrays to have high polarisation purity, no blind spots and no grating lobes. (a) Array of non-radiative dielectric waveguide guide antennas with asymmetric air gaps. (b) Array of printed circuit groove-guide antennas with offset apertures [31]

As each line width is now less than $\lambda_0/2$, no grating lobes will be encountered during the scans, and cross polarised radiation will be suppressed provided the baffles are high enough ($> \lambda_0/2$) so that the mode with vertical polarisation (which is below the cut-off in the air-filled region) has decayed sufficiently at the aperture.

In addition to their excellent electrical performance, leaky-wave arrays have a simple structure, and constitute low-profile antennas that provide design flexibility. They are well suited for a variety of millimetre wave applications. The design of feed systems that match the structural simplicity and electrical performance of the arrays does however require attention.

The scan range θ of the array (see Figure 7.15) measured from the broadside is given by [32]:

$$\sin \theta \approx \frac{\lambda_0}{\lambda_g} - \frac{\lambda_0}{d}$$

where d is perturbation spacing, λ_0 is the free space wavelength, and λ_g is the guided wavelength inside the dielectric rod. A negative angle corresponds to a beam radiated in the backward direction. The radiation angle varies with frequency, and thus the antennas can be used for frequency scanning.

A specific advantage of these antennas is their compatibility with the waveguides from which they are derived, thus facilitating integrated designs. The dielectric grating antennas of Figure 7.15 belong to this class of radiating structures.

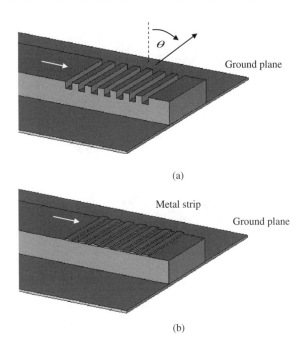

(a)

(b)

Figure 7.15 Periodic dielectric antennas: (a) with dielectric grating and (b) with metal grating. (Reproduced by permission of © 1992 IEEE [33])

7.4 Mutual Coupling between Antennas

In an antenna array, mutual coupling between elements should be taken into account if the antennas are near each other. This section discusses some aspects of the printed antenna elements in an array configuration – in particular, mutual coupling between array elements and array effects on surface wave power. A simple model for mutual coupling is considered to identify the major features of electromagnetic interaction.

Assuming there are two parallel linear dipoles, as shown in Figure 7.16, with the distance between the dipoles being d. If antenna 1 is driven by the input current I_1 and antenna 2 is open-circuited, the near-field generated by the current on antenna 1 will cause an induced voltage V_{21} on antenna 2. The mutual impedance of antenna 2 due to antenna 1 is defined as

Figure 7.16 Parallel dipoles

$Z_{21} = V_{21}/I_{1.}$ If both dipoles are driven, the relationship of the driving voltages to the input currents is given by:

$$V_1 = Z_{11}I_1 + Z_{12}I_2$$
$$V_2 = Z_{21}I_1 + Z_{22}I_2$$

For N parallel dipoles, the driving voltages is:

$$V_p = \sum_{q=1}^{N} Z_{pq}I_q, \quad p = 1, 2, \ldots, K \tag{7.3}$$

When printed antenna elements form an array, mutual coupling levels could be large enough to degrade sidelobe levels and main beam shape, and cause array blindness. These effects can be minimised with the knowledge of coupling between the array elements and its proper inclusion in the array design procedure.

The mutual coupling can be computed as two-port transfer impedance by the moment method. The method yields magnitudes as well as phase, and comparisons with measured data for patches are shown in Reference [34]. A description of the mutual coupling between parallel and collinear half-wave dipoles versus separation is presented in Reference [35].

Figure 7.17 shows the coupling between parallel patches and collinear patches on a ceramic substrate with different substrate thicknesses. The elements are resonant, and the spacing between the elements is a half-wavelength. For thin substrates the coupling levels are very low but increase rapidly with increasing thickness, and then show a tendency to oscillate for thicknesses greater than a half-wavelength.

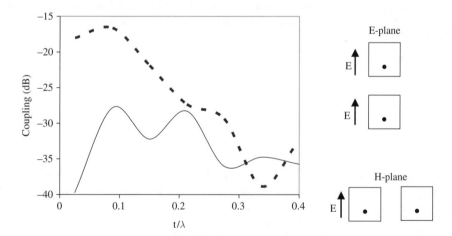

Figure 7.17 Calculated E-plane (solid line) and H-plane (dashed line) mutual coupling magnitude between two microstrip patches with different thicknesses t of a ceramic substrate ($\varepsilon_r = 10$) [14]

The dominant coupling mechanism for the parallel configuration is through *space-wave* fields; since these fields are stronger in the broadside than in the endfire directions, the coupling levels (due to flux linkage) between parallel dipoles are fairly large for close spacings (this may be understood by considering the expressions for the mutual inductance of two parallel conductors), but drop off quickly as the spacing increases. *Surface waves* are launched in the dipole's endfire direction and so have most effect in the collinear configuration.

When elements are combined in an array, the array efficiency changes according to the element excitation, with respect to the efficiency of an isolated element. An example is now given for two collinear half-wave dipoles spaced $\lambda_0/2$ apart on a ceramic substrate. The procedure can be applied to arrays with more than two elements.

As all mutual coupling terms between array elements are included, the overall array efficiency, [14] based on power lost to surface waves, can be defined as the ratio of the total radiated power P_{rad} to the total input power input P_{input}, and can be written as:

$$e = \frac{P_{rad}}{P_{input}} = \frac{[I^*]^t \, \mathrm{Re}\,[Z_{rad}]\,[I]}{[I^*]^t \, \mathrm{Re}\,[Z]\,[I]} \tag{7.4}$$

where $[Z]$ is the total impedance matrix, $[Z_{rad}]$ is the contribution to the radiated field, $[I]$ is a column vector of expansion mode currents and the superscript t denotes the transpose operator. Letting $[R] = \mathrm{Re}[Z]$ and $[R_{rad}] = \mathrm{Re}[Z_{rad}]$, then Equation (7.4) can be written as:

$$e = \frac{[I^*]^t \, R_{rad}\,[I]}{[I^*]^t \, R\,[I]} \tag{7.5}$$

To illustrate the method, Pozar [14] considers an example of two printed dipoles, with three expansion modes on each dipole. The $[Z]$ and $[R]$ matrices are then 6×6 and $[I]$ is a six-element column vector. If the expansion modes are numbered consecutively down each centre-fed dipole, the terminal currents for the first and second dipoles will be I_2 and I_5 respectively. If V_1 is the input voltage applied to dipole 1 and V_2 is the input voltage applied to dipole 2, then:

$$\begin{bmatrix} 0 \\ V_1 \\ 0 \\ 0 \\ V_2 \\ 0 \end{bmatrix} = [Z][I] \tag{7.6}$$

Letting $[I_{10}] = [I]$ for $V_1 = 1$, $V_2 = 0$, and let $[I_{01}] = [I]$ for $V_1 = 0$, $V_2 = 1$. Then by superposition, the dipole currents caused by the port excitation voltages V_1 and V_2 can be simplified as:

$$[I] = [S]\,[V_p], \quad p = 1, 2 \tag{7.7}$$

where $[V_p]$ is a two-element port voltage vector, and $[S] = [[I_{10}], [I_{01}]]$ is a 6×2 matrix. The array efficiency can then be written in terms of the port voltages as:

$$e = \frac{[V_p^*]\,[S^*]^t\,[R_{rad}]\,[S]\,[V_p]}{[V_p^*]^t\,[S^*]^t\,[R]\,[S]\,[V_p]} \tag{7.8}$$

It is noted that Equation (7.8) describes the performance index e, the efficiency, as a ratio of two quadratic forms. Thus, Equation (7.8) can be optimised by solving the eigenvalue equation:

$$[M][V_p] = e[N][V_p] \tag{7.9}$$

where $[M] = [S^*]^t[R_{rad}][S]$, and $[N] = [S^*]^t[R][S]$ are Hermitian 2×2 matrices. A similar procedure has been used for free space array optimisation [36]. If both dipoles are identical then, by symmetry, it can be shown that $[M]$ and $[N]$ have the form:

$$[M] = \begin{bmatrix} m_1 & m_2 \\ m_2 & m_1 \end{bmatrix} \text{ and } [N] = \begin{bmatrix} n_1 & n_2 \\ n_2 & n_1 \end{bmatrix} \tag{7.10}$$

with m_1, m_2, n_1 and n_2 real. The eigenvectors, representing the feed voltages for optimum e, are either even or odd:

$$[V_p]_{\text{even}} = \begin{bmatrix} 1 \\ 1 \end{bmatrix}$$

$$[V_p]_{\text{odd}} = \begin{bmatrix} 1 \\ -1 \end{bmatrix} \tag{7.11}$$

The corresponding eigenvalues are then the efficiencies resulting from the above excitations:

$$e_{\text{even}} = \frac{m_1 + m_2}{n_1 + n_2}$$

$$e_{\text{odd}} = \frac{m_1 - m_2}{n_1 - n_2} \tag{7.12}$$

In general, the even mode produces the maximum array efficiency whereas the odd mode produces the minimum array efficiency. The optimised efficiency e for two collinear half-wave dipoles $\lambda_0/2$ apart with different substrate thicknesses is shown in Figure 7.18, for even and odd mode excitations. As can be seen, the efficiency of the even mode can be improved by as much as 40 % compared to odd mode excitation. Pozar cites a similar calculation for two parallel dipoles that shows a 10 % improvement for even mode excitation.

For the data shown in Figure 7.18, the maximum improvement occurs for uniform phase excitation of the array elements – which is a practical result for broadside arrays. Odd mode

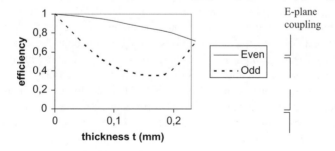

Figure 7.18 Optimised efficiency e for two collinear half-wave dipoles with $\lambda_0/2$ apart versus the ceramic substrate thickness t. (Reproduced by permission of © 1983 IEEE [14])

excitation generally produces a reduced efficiency, which means that more power is being coupled to surface waves – a result that may be of interest for surface wave antennas. In Figure 7.18, the e_{even} and e_{odd} curves cross at the substrate thickness at about $t = 0.22$ mm.

The efficiency variation for planar antenna elements in an array can be partially explained in terms of the phasing of the surface wave fields. Surface waves, launched endfire-wise from each dipole, are significantly out-of-phase (because of the $\lambda_0/2$ dipole spacing) and tend to cancel each other out. It is postulated that element spacing exists such that the maximum cancellation occurs, and the efficiency e approaches 100 % (at least for substrate thicknesses where only one surface wave mode presents). Nevertheless, this situation is unlikely to happen when there is more than one surface wave mode, as the different phase constants would preclude total cancellation.

In addition to the signal coupling between antennas, noise could also couple between elements. Recent findings show that the signal-to-noise ratio is affected by mutual coupling, this points to a new issue about the thermal noise behaviour in a multiantenna system. This section only concentrates on the thermal noise effect. Other noise effects of concern to an antenna array in a communication system can be found in References [37] and [38].

Thermal noise is produced by the random, excited thermally, motion of electrons in a material. The performance of a communication system can be defined in terms of a suitable noise temperature value, T_N, chosen so that:

$$\frac{S}{N} = \frac{T}{T_N}$$

In effect, T_N is the temperature of a thermal source, which would provide a signal power equal to the noise power level. For example, for single-sideband superheterodyne receivers, the noise temperature will be:

$$T_N \geq 0.048 \times f$$

where T_N is in kelvins and f is in GHz. For 60 GHz, the minimum possible noise temperature is 2.88 K.

Thermal noise plays a key role in MIMO communication systems, which use antenna arrays to increase the communication capacity [34]. High capacity could be achieved in these systems by ensuring independence of the channel matrix coefficients, a condition normally achieved with wide antenna element spacing. However, persistent miniaturisation of subscriber units makes such large separations impossible, and the resulting antenna mutual coupling [39] significantly impacts on the communication system performance. In Reference [40], the noise that originated from the amplifier at the receiver end of the MIMO system was included in the consideration, but the thermal noise on the coupled antennas was not considered. The possibility that the thermal noise from a radiating body could be induced in the antenna is discussed in Reference [41]. The topic of partially correlated noise sources that might be introduced into receivers of two closely spaced antennas is discussed in Reference [42].

The impact of antenna mutual coupling on the MIMO system has been evaluated by examining how the coupled antennas change the signal correlation [40]. The modifications in channel matrix coefficients are then used to assess mutual coupling effects on the system capacity [43]. Also the radiated power at the transmitter and the power collection capability due to the effect of this mutual coupling in the multiantenna systems, are presented in Reference [44]. The effect of mutual coupling on the MIMO channel capacity in the context of the signal-to-noise ratio (SNR) is presented in Reference [45].

Thermal noise that originates from the antenna material itself is *self-noise* or *self-radiation*. As well as this self-radiated noise, induced thermal noise appears in the antenna from radiated bodies in the antenna's vicinity [46].

Thermal noise correlation due to mutual coupling effects in closely spaced antennas was missing in the assessment of early MIMO system communication performance with small antenna element separation, which was especially critical for customer units. However, the signal-to-noise ratio (SNR) will now be discussed for the multiantenna system case, with a large number of antennas placed in an infinitesimally volume of space, which models the antenna spacing as almost zero (or small compared to the free space wavelength). Without considering the mutual coupling effect a calculation would give an infinite value for SNR. The result obtained by using classical methods demonstrates the importance of the proper consideration of coupled antennas in multiantenna systems. The expression for the total signal-to-noise ratio is:

$$\text{SNR} = \lim_{\substack{n_R \to \infty \\ d \to 0}} \frac{n_R^2 P}{n_R N} \tag{7.13}$$

where d is the antenna spacing, and P and N are the signal and noise power, respectively.

The multiantenna system can be represented as a general linear network using a generalised form of Thévenin's theorem. The generalisation of the theorem holds true not only for coherent sources but also for thermal noise sources [47]. It is valid even for a general linear network that may contain a number of inaccessible (hidden) nodes together with internal voltage and current sources, whose location may be unknown. However, as long as there are only N independent accessible nodes, such a system is indistinguishable from a noise source free network, with the same impedance or admittance matrix, together with a set of N nodal current generators of infinite internal impedance. The current from the generator of the ith node, in such an equivalent network, is equal to the current flowing into the rth node of the original network when all nodes of the latter are short-circuited to earth. The internal sources may be alternatively represented by a set of N nodal voltage generators of infinite internal admittance, such that the voltage across the generators in the rth node is equal to the voltage across the rth node of the original network; when all the nodes of the latter are open-circuit. The nodal noise sources are not in general independent.

The multiantenna system with $N = n_R$ antenna elements, can be represented as a linear n_R terminal-pair network containing internal signals or noise generators; it is specified completely with respect to its terminal pairs by its admittance matrix $\mathbf{Y}$ and a set of n_R nodal current generators $i_1, i_2, \ldots, i_{n_R}$. In matrix form, $\mathbf{Y}$ denotes a squared matrix of order n_R:

$$\mathbf{Y} = \begin{pmatrix} y_{11} & y_{12} & \cdots & y_{1n_R} \\ y_{21} & y_{22} & \cdots & y_{2n_R} \\ \cdots & \cdots & \cdots & \cdots \\ y_{n_R 1} & y_{n_R 2} & \cdots & y_{n_R n_R} \end{pmatrix} \tag{7.14}$$

The complex amplitudes of thermal current generators are represented conveniently by a column vector i:

$$i = \begin{pmatrix} i_1 \\ i_2 \\ \cdot \\ i_{n_R} \end{pmatrix} \tag{7.15}$$

The nodal noise sources are not in general independent. Therefore the spectral density of the squared current can be written in matrix form as:

$$
\overline{\vec{u}\vec{u}^+} = \begin{pmatrix} \overline{i_1 i_1^+} & \overline{i_1 i_2^+} & \cdots & \overline{i_1 i_{n_R}}^+ \\ \overline{i_2 i_1^+} & \overline{i_2 i_2^+} & \cdots & \overline{i_2 i_{n_R}}^+ \\ & & & \\ \overline{i_{n_R} i_1^+} & \overline{i_{n_R} i_2}^+ & \cdots & \overline{i_{n_R} i_{n_R}}^+ \end{pmatrix}
\tag{7.16}
$$

where the subscript $^+$ indicates the Hermitian transpose (complex conjugate transpose).

The isolated receivers of two closely spaced antennas will receive partially correlated noise [48]. The magnitude correlation was calculated using a generalised form of Nyquist's thermal noise theorem, given in Reference [49]. It was shown that a general non-reciprocal network with a system of internal thermal generators, all at temperature T, is equivalent to the source-free network together with a system of noise current generators I_r and I_s and with infinite internal impedance [50]. Noise currents are correlated and their cross-correlation is given by:

$$
\overline{I_S I_T} \mathrm{d}f = 2kT(Y_{ST} + Y_{ST}^*)\mathrm{d}f
\tag{7.17}
$$

where Y_{ST} is the mutual admittance.

Writing the internal noise sources as a system of nodal voltage generators V_T and V_s, with zero internal impedance. The correlation of nodal voltage generators is then given by:

$$
\overline{V_S V_T} \mathrm{d}f = 2kT(Z_{ST} + Z_{ST}^*)\mathrm{d}f
\tag{7.18}
$$

where Z_{ST} is the mutual impedance. The correlation is zero when the mutual coupling is purely reactive.

The application of the generalised Nyquist thermal noise theorem allows us to determine thermal noise power of coupled antennas in the multiantenna system. The theorem states that for a passive network in thermal equilibrium it would appear to be possible, to represent the complete thermal noise behaviour by applying Nyquist's theorem independently to each component element of the network. Coupling of the multiantenna system is represented by antenna self-impedances and mutual impedances. In order to determine thermal noise behaviour, self-impedances as well as mutual impedances should be taken into account. The thermal noise power calculation, which implies a mutual coupling effect, is given for the multielement array. The generalisation for the multiantenna system can consequently be made.

$$
\mathbf{I} = \begin{bmatrix} i_{L1} + i_1 \\ i_{L2} + i_1 \\ . \\ i_{Ln_R} + i_{n_R} \end{bmatrix}
\tag{7.19}
$$

The nodal network representation for the multiantenna system with n_R antenna elements is shown in Figure 7.19:

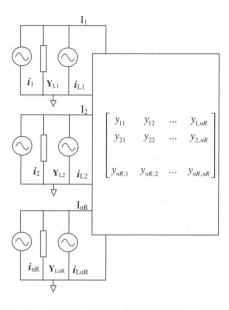

Figure 7.19 Nodal network representation for a multiantenna system. (Reproduced by permission of © 2005 IEEE [49])

The admittance and voltage matrix of this system can be written as:

$$\mathbf{Y} + \mathbf{Y}_L \mathbf{U} = \begin{bmatrix} y_{11} + \mathbf{Y}_L & y_{12} & \cdot & y_{1n_R} \\ y_{21} & y_{22} + \mathbf{Y}_L & \cdot & y_{2n_R} \\ \cdot & \cdot & \cdot & \cdot \\ y_{n_R 1} & y_{n_R 2} & \cdot & y_{n_R n_R} + \mathbf{Y}_L \end{bmatrix} \tag{7.20}$$

$$\mathbf{V} = \begin{bmatrix} V_1 \\ V_2 \\ \cdot \\ V_{n_R} \end{bmatrix} \tag{7.21}$$

Thus the system in Figure 7.19 can be expressed as follows:

$$\mathbf{I} = \mathbf{i} + \mathbf{i}_L = (\mathbf{Y} + \mathbf{Y}_L \mathbf{U})\mathbf{V} \tag{7.22}$$

where $\mathbf{U}$ is a unitary vector. Equation (7.22) can be rewritten as:

$$\mathbf{V} = (\mathbf{Y} + \mathbf{Y}_L \mathbf{U})^{-1}\mathbf{I} = (\mathbf{Y} + \mathbf{Y}_L \mathbf{U})^{-1}(\mathbf{i} + \mathbf{i}_L) \tag{7.23}$$

$$N = \frac{1}{2}(\mathbf{Y}_L + \mathbf{Y}_L^*)\overline{\mathbf{V}\mathbf{V}^+} \tag{7.24}$$

$$\mathbf{V}\mathbf{V}^+ = (\mathbf{Y} + \mathbf{Y}_L \mathbf{U})^{-1}\mathbf{I}\mathbf{I}^+((\mathbf{Y} + \mathbf{Y}_L \mathbf{U})^{-1})^+ \tag{7.25}$$

where the symbol $^+$ indicates the Hermitian transpose and * the complex conjugate. Then, the square of the current can be expressed as:

$$\mathbf{II}^+ = (\boldsymbol{i} + \boldsymbol{i}_L) \times (\boldsymbol{i} + \boldsymbol{i}_L)^+ \tag{7.26}$$

Based on Equation (7.17) the following relations are valid and can be put into Equation (7.26):

1. $$\overline{i_j i_k^*} = 2kT(y_{jk} + y_{jk}^*)$$

2. $$\overline{i_{Lj} i_k^*} = 0 \tag{7.27}$$

3. $$\overline{i_{Lj} i_{Lk}^*} = 0, \quad j \neq k$$

Finally, the square of the currents (7.26) is expressed as:

$$\mathbf{II}^+ = 2kT(\mathbf{Y} + \mathbf{Y}^* + (\mathbf{Y}_L + \mathbf{Y}_L^*)^* \mathbf{U}) \tag{7.28}$$

Substituting $\mathbf{Y}_L + \mathbf{Y}_L^* = 2G_L$ and $\mathbf{Y}_a = \mathbf{Y} + \mathbf{Y}_L \mathbf{U}$, Equation (7.24) can be written as

$$N = 2kTG_L \left[\mathbf{Y}_a^{-1} \times (\mathbf{Y}_a + \mathbf{Y}_a^*) \times (\mathbf{Y}_a^{-1})^+ \right] \tag{7.29}$$

The thermal noise received in each antenna element includes two components, self-thermal noise and induced thermal noise from the adjacent antenna elements. The total thermal noise power received from the antenna array in the receiver load is given in Equation (7.29). The total noise for two coupled antenna elements, in the frequency bandwidth B, can be simplified to be a sum of these noise powers:

$$N_{\text{total}} = \int_B P_{L1} \mathrm{d}f + \int_B P_{L2} \mathrm{d}f \tag{7.30}$$

where P_{L1} is the thermal noise power absorbed in the receiver load of the first antenna, and P_{L2} is the thermal noise power absorbed in the receiver load of the second antenna.

To minimise mutual coupling between elements in a printed substrate, it is possible to create additional isolation between antenna elements, such as designing slots as in Figure 7.19, or adding absorbing material. It is found that elements with better isolation can prevent the distortion of the radiation pattern and thus increase the output gain [51].

References

[1] Tolga M. Duman and Ali Ghrayeb, '*Coding for MIMO Communication Systems*', John Wiley & Sons, Ltd, Chichester, January 2008.

[2] William Webb, '*Wireless Communications: The Future*', John Wiley & Sons, Ltd, Chichester, March 2007.

[3] Athos Kasapi, "http://www.bookfinder.com/dir/i/Smart_Antennas_and_Adaptive_Arrays-Multi-Antenna_ Techniques_for_Wireless/0750678097/ *Smart Antennas and Adaptive Arrays: Multi-Antenna Techniques for Wireless Communications*" Butterworth-Heinemann Limited, ISBN 0750678097 (0-7506-7809-7), 2008.

[4] K. Huang and D. J. Edwards, '60 GHz Multi-beam Antenna Array for Gigabit Wireless Communication Networks', *IEEE Transactions on Antennas and Propagation*, **54**(12), December 2006, 3912–3914.

[5] M. Uno, Z. Wang, V. Wullich and K. Huang, 'Communication System and Method', Patents US2006116092, EP1659813 and JP2006148928, May 2006.

[6] S. Kobayashi, R. Mittra and R. Lampe, 'Dielectric Tapered Rod Antenna for Millimetre Wave Applications', IEEE Transactions on Antennas and Propagation, **30**(1), January 1982, 54–58.

[7] J. Kraus and R. Marhefka, *'Antennas for All Applications'*, 3rd edition, McGraw-Hill, New York, 2002.

[8] T. Ando, J. Yamauchi and H. Nakano, 'Demonstration of the Discontinuity-Radiation Concept for a Dielectric Rod Antenna', Proceedings of IEEE Antennas and Propagation Society International Symposium, 16–21July 2000, Vol. 2, pp. 856–859.

[9] K. Huang and Z. Wang, 'V-Band Patch-Fed Rod Antennas for High Datarate Wireless Communications', *IEEE Transactions on Antennas and Propagation*, **54**(1), January 2006, 297–300.

[10] K. Huang and Z. Wang, 'Dielectric Rod Antenna and Method for Operating the Antenna', Patents WO2006097145 and EP1703590, 21 September 2006.

[11] IEEE 802.15.3c Standard, http://www.ieee.org/

[12] T. Ando, J. Yamauchi and H. Nakano, 'Rectangular Dielectric-Rod by Metallic Waveguide', *IEE Proceedings on Microwave, Antennas and Propagation*, **149**(2), April 2002, 92–97.

[13] Kai Chang, *'Phased Array Antennas'*, John Wiley & Sons, Inc., New York, 2001.

[14] David M. Pozar, 'Considerations for Millimetre Wave Printed Antennas', *IEEE Transactions on Antennas and Propagation*, **AP-31**(5), September 1983, 740–747.

[15] M. A. Weiss, 'Microstrip Antennas for Millimetre Waves', *IEEE Transactions on Antennas and Propagation*, **AP-29**, January 1981, 171–174.

[16] D. M. Pozar and D. H. Schaubert, 'Comparison of Architectures for Monolithic Phased Array Antennas', *Microwave Journal*, **29**, March 1986, 93–104.

[17] J. R. James and C. M. Hall, 'Investigation of New Concepts for Designing Millimetre-Wave Antennas', Final Technical Report on Contract DAJA37-80-C-0183, US Army European Research Office, September 1983.

[18] P. B. Katehi and N. G. Alexopoulos, 'On the Modeling of Electromagnetically Coupled Microstrip Antennas – The Printed Strip Dipole', *IEEE Transactions on Antennas and Propagation*, **AP-32**, November 1984, 1179–1186.

[19] P. B. Katehi, N. G. Alexopoulos and I. Y. Hsia, 'A Bandwidth Enhancement Method for Microstrip Antennas', *IEEE Transactions on Antennas and Propagation*, **AP-35**, January 1987, 5–12.

[20] R. J. Mailloux, *'Phased Array Antenna Handbook'*, Artech House Inc, Norwood, Massachusetts, 2005.

[21] R. J. Mailloux, 'Phased Array Architecture', *IEEE Proceedings*, **80**(1), January 1992, 163–172.

[22] A. K. Agrawal and E. L. Holzman, 'Beamformer Architectures for Active Phased-Array Radar Antennas', *IEEE Transactions on on Antennas and Propagation*, **AP-47**(3), March 1999, 432–442.

[23] B. Rama Rao, '94 GHz Slotted Waveguide Array Fabricated by Photolithographic Techniques', *Electronics Letters*, **20**(4), 16 February 1984, 155–156.

[24] H. Y. Yee, 'Impedance of a Narrow Longitudinal Shunt Slot in a Slotted Waveguide Array', *IEEE Transactions on Antennas and Propagation*, **AP-22**, 1974, 589–592.

[25] A. A. Oliner, 'The Impedance Properties of Narrow Radiating Slots in the Broadface of Rectangular Waveguide', *IEEE Transactions on Antennas and Propagation*, **AP-5**, 1957, 12–20.

[26] U. Kotthaus and B. Vowinkel, 'Investigation of Planar for Submillimeter Receivers', *Transactions of the IEEE*, **MTT-37**, February 1989, 375–380.

[27] G. M. Rebeiz and D. B. Rutledge, 'Integrated Horn Antennas for Millimetre-Wave Applications', *Annals of Telecommunications*, **47**, 1992, 38–48.

[28] Y. Huang, 'Integrated mm-Wave Planar Array Antenna with Low Loss Feeding Network', World Patent WO 2006/097050 A1.

[29] K. Noujeim and K. Balmain, 'Fixed-Frequency Beam-Steerable Leaky-Wave Antennas', PhD Dissertation, Department of Electrical Computational Engineering, University of Toronto, Toronto, Ontario, Canada, 1998.

[30] M. Guglielmi and A. A. Oliner, 'A Practical Theory for Image Guide Leaky-Wave Antennas Loaded by Periodic Metal Strips', Proceedings of the 17th European Microwave Conference, Rome, Italy, 11–17 September 1987, pp. 549–554.

[31] A. A. Oliner, 'Scannable Millimetre Wave Arrays', Weber Research Institute, Polytechnic University, Technical Report Poly-WRI-1543-88, Vols I and 11, 30 September 1988.

[32] Limin Huang, Jung-Chih Chiao and Michael P. De Lisio, 'An Electronically Switchable Leaky Wave Antenna', *IEEE Transactions on Antennas and Propagation*, **48**(11), November 2000, 1769–1772.

[33] Felix K. Schwering, 'Millimetre Wave Antennas', *Proceedings of the IEEE*, **80**(1), January 1992, 92–102.

[34] D. M. Pozar, 'Input Impedance and Mutual Coupling of Rectangular Microstrip Antennas', *IEEE Transactions on Antennas and Propagation*, **AP-30**, November 1982, 1191–1196.

[35] N. G. Alexopoulos and I. E. Rana, 'Mutual Impedance Computation between Printed Dipoles', *IEEE Transactions on Antennas and Propagation*, **AP-29**, January 1981, 110–111.

[36] R. F. Hamington, *'Field Computation by Moment Methods'*, Macmillan, New York, 1968.

[37] Michael J. Gans, 'Channel Capacity between Antenna Arrays – Part I: Sky Noise Dominates', *IEEE Transactions on Communications*, **54**(9), September 2006, 1586–1592.

[38] Michael J. Gans, 'Channel Capacity between Antenna Arrays – Part II: Amplifier Noise Dominates', *IEEE Transactions on Communications*, **54**(11), November 2006, 1983–1992.

[39] I. J. Guptha and A. K. Ksienski, 'Effect of the Mutual Coupling on the Performance of the Adaptive Arrays', *IEEE Transactions on Antennas and Propagation*, **31**(5), 1983, 785–791.

[40] W. Rotman, 'EHF Dielectric Lens Antenna for Multibeam MIL-SATCOM Applications' Digest of 1982 International IEEE-APSIURSI Symposium, Albuquerque, New Mexico, June 1982, pp. 132–135.

[41] K. Iizuka, M. Mizusawa, S. Urasaki and J. Ushigome, 'Volume-Type Holographic Antenna', *IEEE Transactions on Antennas and Propagatation*, **AP-23**, November 1975, 807–810.

[42] P. Bhartia, K. V. S. Rao and R. S. Tomar, *'Millimetre-Wave Microstrip and Printed Circuit Antennas'*, Artech House, London, 1991.

[43] T. Sventenson and A. Ranheim, 'Mutual Coupling Effects on the Capacity of the Multielement Antenna System', Proceedings of the International Conference on *'Acoustic, Speech and Signal Processing'* (ICASSP'01), 2001.

[44] J. W. Wallace and M. A. Jensen, 'Mutual Coupling in MIMO Wireless Systems: A Rigorous Network Theory Analysis', *IEEE Transactions on Wireless Communications*, **3**, 2004, 1317–1325.

[45] S. Krusevac, P. Rapajic and R. Kennedy, 'Method for MIMO Channel Capacity Estimation for Electro-magnetically Coupled Transmit Antenna Elements', AusCTW 2004 , February 2004, pp. 122–126.

[46] S. M. Rytov, Yu. A. Krastov and V. I. Tatarskii, *'Principles of Statistical Radiophysics 3 , Elements of the Random Fields'*, Springer, Berlin, Heidenberg, New York, 1987.

[47] A. T. Starr, *'Electric Circuit and Wave Filters'*, 2nd edition, Pitman, London, 1946, p. 78.

[48] S. Krusevac, P. B. Rapajic, R. A. Kennedy and P. Sadeghi, 'Mutual Coupling Effect on Thermal Noise in Multi-antenna Wireless Communication Systems', Proceedings of the 6th Australian Communications Theory Workshop, February 2005, pp. 209–214.

[49] G. E. Vally and H. Wallman, *'Vacuum Tube Amplifiers'*, MIT Radio Laboratory Series 18, McGraw-Hill, Inc., New York, 1949.

[50] R. Q. Twiss, 'Nyquist's and Thevenin's Generalized for Nonreciprocal Linear Networks', *Journal of Applied Physics*, **26**, May 1955, 559–602.

[51] Y. Moon *et al.*, 'Flat-Plate MIMO Array Antenna with Isolation Element', US Patent Application US2007/0069960.

8

Smart Antennas

In an indoor environment, millimetre wave communications can have both line-of-sight and non-line-of-sight wireless links [1], and it is difficult for an omnidirectional or fixed beam antenna to cope with both situations. A *smart* antenna system, which is an antenna array arranged in a special distributed configuration with a specialised signal processor, can be deployed in a millimetre wave communication system to dynamically optimise the system's performance and capacity significantly by minimising undesired co-channel interference. Therefore, smart antennas are attractive as they can help to improve the millimetre wave communication quality. Various smart antenna technologies such as spatial diversity combining or beamforming algorithms can be implemented to enhance the performance of a millimetre wave system. Most of the smart antenna systems published can form beams directed to a desired signal and form nulls towards an undesired interferer, such as a co-channel base station. This enhances the signal-to-interference (SIR) ratio because the received desired signal strength is maximised and the undesired signal interference is minimised. The other benefits of a smart antenna are as follows:

- Increase in range or coverage arising from an increased signal strength due to array gain
- Increase in capacity arising from interference rejection
- Reject multipath interference arising from inherent spatial diversity of the array
- Reducing expense arising from lower transmission powers to the intended end user

Smart antenna systems are commonly classified into several categories: mode tracking, beam switching, beam steering/forming and multiple-input multiple-output. They can isolate a co-channel signal whereas multiple antennas cannot.

A beam-switching antenna system consists of many highly directive, pre-defined fixed beams formed with an antenna array. The system usually detects the maximum received signal strength from the antenna beams, and chooses to transmit the output signal from one of the selected beams that gives the best performance. In many ways, a nine-beam switched antenna system is very much like an extension of a sectoring directional antenna with multiple subsectors. Since the direction of the arrival information of the desired signal is not analysed, the desired signal may not fall on to the maximum of the chosen beam because the direction of these beams is fixed. Therefore switched-beam antennas may not provide the optimum SIR. In fact, in cases where a strong interfering signal is at or near the centre of the chosen beam, while the desired user is away from the centre of the chosen beam, the interfering signal can be enhanced far more than the desired signal, which results in very poor SIR.

Beam-steering antennas use the direction of arrival (DoA) information from the desired signal and steer a beam maximum towards the desired signal direction. This method, when compared to that of the switched-beam antennas, is far superior in performance because it tracks the desired user direction of arrival continuously, using a tracking algorithm to steer the beam towards the desired user. There is a considerable amount of published work that addresses the problem of sensing the DoA and indeed of sensing the DoA of an interferer [2]. Real-time beamforming (typically for array antennas) and signal tracking, enables the system (depending on the details of the beam-steering/forming algorithm) to maintain optimal gain for the desired signal, while simultaneously minimising the reception of the interfering signal by directing the nulls to an interferer's direction. Thus excellent SIR can be achieved. For reflector antennas, the problem of optimising beam pointing (source tracking) in real time is approached in a different way. Often these antennas are served by one feed and the whole assembly is mechanically moved to point the beam to the antenna in the direction of the target. In complex reflector antennas (for instance, providing shaped-beam coverage) the feed consists of an array of horns. In either case the source tracking can be implemented by control of the aperture distribution of the feed; to directly provide real-time pointing control or as a means of direction finding, by temporarily deviating the antenna pointing to ascertain the direction of arrival of the signal. The control systems that ultimately determine the pointing of the antenna can be divided into three classes.

The first is a full real-time closed-loop control technique, which utilises the aperture distribution of the feed or the modal excitation within the feed structure to detect phase and amplitude errors (which generate higher-order modes in a waveguide) in the incident wave front, and uses this information to drive a closed-loop control to null out these errors. Such systems are generically called *monopulse* systems, a term derived from early radar. In a waveguide feed mode, couplers placed near the throat of a feed horn detect the phase and amplitude of higher-order modes and deliver this information to the (analogue) control system. Clearly, this can be an expensive (although very accurate) system, which requires several coherent receiver systems. A cheaper approach used in early radar was to scan the main beam conically around the nominal pointing of the antenna and use the observed amplitude modulation in a control loop; the antenna was pointed at the source when no modulation was observed. Further advances appeared in the early 1970s when microprocessors became available. The analogue scanning was replaced by a stepping and hill-climbing technique in which the pointing of the antenna was stepped from side to side in order to work out where the signal was coming from. The control was simple and cheap. However, these later amplitude-only systems, while being considerably cheaper than monopulse, could not cope with highly dynamic situations in which the source moved quickly or the signal fluctuated within the time scale of the scanning or step cycle. With the advent of more powerful computers and semi-conductor components, the possibility of combining the features of monopulse and step track with more sophisticated control algorithms became viable. The use of a behavioural model of the source enabled performance approaching the monopulse systems but with the economical cost of step trackers, to be established. A horn feed that is able to provide a tracking method for waveguide and horn antennas, based on generating TE_{21} and TM_{01} modes to achieve rapid step-tracking behaviour, will be introduced in Section 8.4.

8.1 Beam-Switching Antennas

A beam-switching antenna consists of generating a multiplicity of juxtaposed beams (generated by an array) whose output may be switched to a receiver or a bank of receivers. The addressed space is therefore served by a set of beams that may be switched on or off according to an algorithm that is able to sense the desired direction of transmission or reception. Beam-switching antennas may, in some cases, be cheaper than an equivalent phased array at millimetre wave communications, particularly when few beams are needed. It deploys a fixed set of relatively narrow azimuthal beams. An example of a low-cost switched-beam array is shown in Figure 8.1.

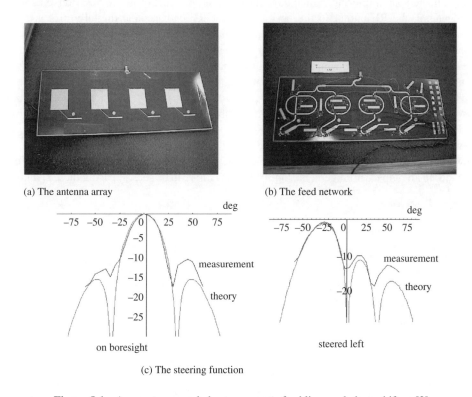

(a) The antenna array

(b) The feed network

on boresight

steered left

(c) The steering function

Figure 8.1 An aperture-coupled antenna array, feed lines and phase shifters [3]

In this antenna, the addressed space is 90° in front of the antenna. The beams are approximately 50° wide and the feed network uses fixed-value switched phase shifters to steer the beam. In this example, amplifier stages are integrated into each feed line to counter the slight loss in the phase shifters. In this configuration the antenna operated in the receive mode only. For combined transmit–receive operation the amplifiers would be omitted. The radiation patterns are shown in Figure 8.1. (The steered-right condition is not included, being a mirror image of the steered-left case.) It can be seen that the antenna gives spatial coverage at $+/-3\,\mathrm{dB}$ over 90° [3, 4].

When signals travel between the transmitter and receiver, they can be reflected, scattered, diffracted and shadowed. As a result, these signals experience fading because of attenuation and phase shifts when they are combined at the receiver. Another source that degrades the performance of signal reception in a mobile environment is interference. Techniques that overcome these impairments and improve system performance are examined in this chapter; namely diversity and adaptive beamforming, both of which are spatial techniques. Diversity techniques can provide a diversity gain or a reduction in the margin required to overcome fading. In a digital communication system, this results in an improvement in the required signal-to-noise ratio (SNR) or the ratio of energy per bit to noise power spectral density (Eb/No) necessary to achieve a given quality of service in terms of the bit error rate (BER). Similarly, beamforming provides several types of improvements in terms of array gain, interference reduction and spatial filtering, which have the cumulative effect of improving Eb/No as well.

In general, the RF output to the beams is either RF or baseband digitally processed to ascertain the sector in which the communicating mobile device may be located. The coverage is broken down into sectors, with each sector served by an array of radiating elements fed by a beam-switching network, which ideally forms independent beams.

In the millimetre wave range, a simple switching network is provided by single-pole double-throw switches together with a corporate feeding network. For a 2×2 array, the switching network is shown in Figure 8.2 (a); and the number of switches is 3. For an $N \times N$ array, it would be necessary to have $2^N + 2^{N-1} + 2^{N-2} + \cdots + 2^1 + 2^0$ switches. The other type of switching network uses single-pole triple-throw switches with a series feeding network, as shown in Figure 8.2 (b). Each switch can select the signal path from three choices: patch, feed and load. In all of these arrangements, one or more beams can be selected using these switches.

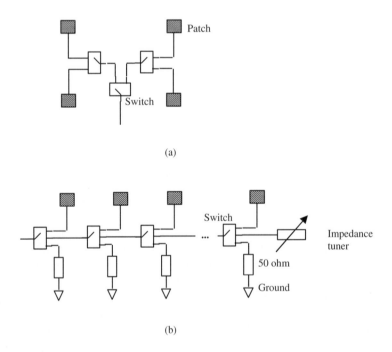

(a)

(b)

Figure 8.2   (a) Corporate feeding network with single-pole double-throw switches. (b) Series feeding network with single-pole triple-throw switches

Generally, millimetre wave switches have a lower cost than millimetre wave phase shifters. Therefore, the implementation cost of a beam-switching array is lower than for the case of other types of smart antennas.

The conventional network used to form the beam-switching function is the Butler matrix [5], which consists of a series of interconnected fixed phase shift sections and 3 dB hybrid couplers. However, the number of inputs and outputs is constrained to powers of two; beam shape control is difficult, and beam scanning with frequency occurs (the phase shifters are essentially single frequency), destroying beam independence and leading to a narrow operational bandwidth.

The arrangement of the Butler matrix for an eight beam system is shown in Figure 8.3. There are eight input ports and eight output ports. This is a reciprocal structure, so either end can be the RF input or RF output. The matrix consists of quad hybrid couplers and phase shifters. The number of each depends on the number of beams generated. For a linear array of N elements, the number of couplers is $(N/2) \log_2 N$, where N is the number of beams and the number of phase shifters is $(N/2)(\log_2 N - 1)$. For millimetre wave personal communications, the number N will be modest. To improve flexibility, the use of six- and eight-port hybrid couplers [6] and termination of unused ports [7] has been suggested. Asymmetric hybrids and orthogonal beam synthesis are also possible.

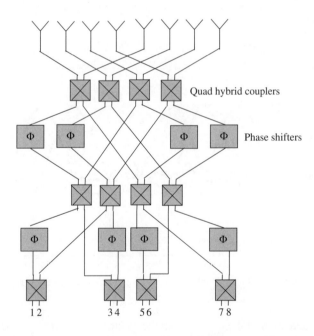

Figure 8.3   A Butler matrix labyrinth producing multiple spot beams

The maximum matrix size is limited by the manufacturing tolerance, which must be less than the smallest phase shift element required. The introduction of 180° hybrids [8] reduces the number of phase shifters significantly; for example, when N = 32, 15 are saved. In addition the minimum phase shift is twice that for a 90° hybrid matrix, and this eases the maximum matrix size constraint. A further reduction in the hybrid count can be achieved using reflective matrices [9, 10].

Since the Butler array is uniformly illuminated, the pattern shape is $\sin(x)/x$ (for a square aperture), the patterns are spaced to be orthogonal and the crossover is 4 dB down, with sidelobes down 13.2 dB [2]. The direction of the beams is dictated by the separation of the antenna elements. A typical array factor for eight elements is shown in Figure 8.4. Its features are as follows:

- The beams are equally spaced and the peaks are located at the nulls of the other beams.
- Since the array is uniformly illuminated, this gives the smallest beamwidth possible with maximum gain. This follows from array theory. The first sidelobes are down 13.2 dB. Each array pattern has the shape $\sin(x)/x$ and the array of beams generated is of the form $\sin(NX)/X$.
- There are scalloped beams with crossover of the beams occurring at the -3.9 dB level.
- The beams are orthogonal and outputs are therefore isolated from each other. Orthogonality also implies that the network is lossless except for the insertion loss. The latter is kept small by judicious circuit design.
- There is no boresight beam. The Bulter matrix, using quadrature hybrids, does not produce a boresight beam. It can be produced if the quadrature hybrids are replaced by hybrid rings or 0–180 hybrids.

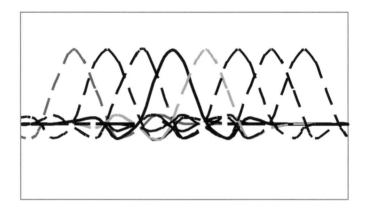

Figure 8.4   Orthogonal beams for an eight-element Butler array

The method of detection of the direction, in the electronic scan method can be either electronic switching of the beam, or use of digital beamforming (DBF) or a monopulse technique to process signals received simultaneously at several receiving antennas [11]. In comparison with the first, the second has the advantages that the scanning range and the step width can be controlled freely. However, the number of receivers increases because one receiver for each of the receiving antennas are usually required. Therefore, a method has also been conceived in which several receiving antennas are controlled by switches so that only one receiver is used [12].

In order to realise simultaneously both a wide field of view and high resolution, the antenna element spacing should be reduced and the antenna aperture should be increased. However, as the antenna size increases the antenna configuration becomes more complex due to an increase in the number of elements. In order to resolve these problems, an electronic scanning system can be designed that is able to perform azimuthal detection by signal processing. By switching not only the receiving antennas but also the transmitting antennas, the number of elements is reduced, thus simplifying the system.

The details of the hardware implementation of these beamforming techniques can vary. Considering the need to avoid expensive phase shifters, beamforming arrays can use quadrature hybrids to generate the real and imaginary components of the signal, and use (real) attenuators to then recombine the components.

In addition, in order to make the antenna small, a resolution-improving technique, such as the estimation of signal parameters via rotational invariance techniques (ESPRIT) [13], can be used instead of the digital beamforming (DBF) used in conventional antennas. This was the approach adopted in an automotive electronic scan millimetre wave antenna [14]. In comparison with DBF, ESPRIT provides much higher resolution [15] so that the aperture size needed for the same azimuthal resolution can be smaller. However, although the ESPRIT algorithm assumes a point source, the target vehicle in this case [15] had multiple reflection points with different intensities distributed over the entire vehicle body [16, 17].

Figure 8.5 depicts the configuration of a typical electronic scan antenna with several receiver channels [18]. In this example the system consists of a transmitting antenna and several receiving antennas placed along a line at equal intervals. Each receiving antenna is connected to an independent receiver. To these receivers, a phase-locked local oscillator signal is applied to a mixer, which produces an output at the baseband. In this configuration, the receiver produces a baseband signal generated by synchronous detection. After analogue to digital conversion, the baseband signal is input

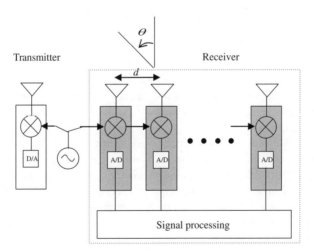

Figure 8.5 Configuration of a fundamental one-dimensional electrically scanned antenna for applying the superresolution technique to detect the angular position of the target [18]

to the signal processing unit, where the direction of arrival of the received wave is estimated.

The direction θ at which the phase is inverted, or where the phase difference of the received signals obtained by adjacent receiving antennas is 180°, is given by the diffraction grating equation:

$$d \sin \theta = \pm \frac{\lambda}{2}$$

where d is the spacing of the receiving antennas. Thus in order to obtain a field of view (the main beam) of approximately $\pm\,20°$, the spacing of the receiving antenna elements must be 1.5 wavelengths. For d equal to 1.5 times the wavelength, $\theta = \pm\,19.5°$. This effective field of view will not generate phase folding (phase ambiguity/repeating or wrap-around) until the view angle is about 39°. If a wave arriving from a direction outside the effective field of view is received, the direction of arrival is ambiguous and falsely calculated. Therefore, it is desirable that the effective field of view be sufficiently wide for this problem to be avoided.

Angular resolution is determined by the antenna beamwidth, which is in turn governed by the aperture size and the transmitter frequency. A rule of thumb is that the 3 dB beamwidth and aperture width (or diameter) W, can be written as [19]:

$$\theta_{3dB} = \frac{70\lambda}{W} \text{ deg}$$

For nine elements at 1.5λ spacing this gives an aperture of 12λ. As the 3 dB beamwidth is about 6–7°, it should be possible to discriminate between two sources spaced at about 2°. It has been shown that an angular resolution (source discrimination) of less than 2° can be attained with a nine-channel electronic scan antenna with nine receiving antennas spaced at 1.5 wavelengths [18].

However, there remain several problems before the realisation of the electronic scan antenna shown in Figure 8.5. For instance, if the phase delay in one of the receivers is different from those in the other receivers, the accuracy and resolution of angle detection by the superresolution method is clearly degraded. There is the probability that the phase delays of nine receivers may fluctuate as a result of temperature variations.

Furthermore, the cost is increased because in the feed network of the local oscillator signal nine receivers are needed. To avoid this problem a configuration has been suggested in which only one receiver is used and the antennas are rapidly switched [20]. In order to switch the nine receiving antennas needed in this antenna, four switches are needed if SP3T (single-pole triple-throw) switches were used.

Figure 8.6 shows the configuration of such an electronic scan antenna with switched transmitting and receiving antennas. This proposal allows for a simpler and smaller electronic scan antenna. In this configuration, three transmitting antennas and three receiving antennas were mounted with different spacings. Both the transmitter and the receiver have SP3T (single pole triple throw) switches for switching three antenna elements. By using two switches, signals equivalent to the nine-channel electronic scan antenna can be obtained by time division multiplexing. It is then possible to reduce the number of antenna elements from Figure 8.5 to those in Figure 8.6. However, due to the transmission loss in the switches, the SNR is degraded and the angular resolution may be decreased. Thus, it is important to minimise the transmission loss in millimetre wave switches.

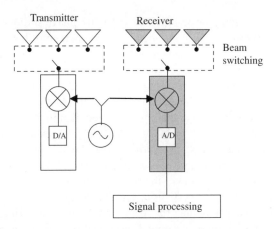

Figure 8.6 Configuration of new electrically scanned antenna with switching of transmitting and receiving antennas [18]

8.2 Beam-Steering/Forming Antennas

Beam steering is a user-specific beamforming method, where each user is served with an individual beam. In the multiple fixed-beam method in Section 8.1, users are served with the beam with the lowest path loss. Beam steering produces a unique beam for each user in order to transmit the signal for a user only into the direction where the signal experiences the lowest path loss as it travels to the user while simultaneously keeping the transmit power into other directions as low as possible.

However, this only applies to dedicated traffic channels, although the common signalling channels still have to be transmitted to the entire sector with a single antenna element of the antenna array. Figure 8.7 illustrates the equipment needed for user-specific beamforming, including:

- A signal processing unit capable of serving, calculating and applying appropriate antenna weights for all users, plus common channels sent/received, into/from the entire sector
- An antenna array
- A power amplifier per antenna element of the antenna array

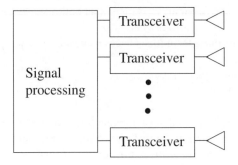

Figure 8.7 Beamforming configuration

The beam-steering function can be operated either by phase shifters at millimetre wave frequencies or by signal processing at baseband. The advantage of beam steering is that the transmit power is essentially concentrated towards the desired user(s). Therefore, beam steering should provide a higher capacity gain than the fixed-beam methods. In addition, as the serving beam tracks the mobile devices, there is no need to hand over the mobile from beam to beam while it is travelling through the coverage area of the sector, as would be necessary in fixed-beam schemes. Hence, far less signalling is needed than in fixed-beam methods, where signalling is necessary every time a mobile moves from the coverage area of one beam to another. The main disadvantage of beam steering is that the base station has to determine the optimum transmit direction for all active users in the sector, which is computationally very intensive compared with determination of the serving beam in the fixed-beam method.

The wide range of multiple-beam systems means that developments in this technology have themselves been diverse, with techniques for operating at both radio and intermediate frequency (IF), in addition to methods employing digital or optical frequency methods. IF, digital and optical beamforming are major topics in their own right and are not dealt with here. Similarly, radio frequency (RF) beamforming is a large topic, one that not only presents a bewildering variety of types but also leads, in some cases, to uncertainty about the best technique to be applied to a given problem. The next section aims to introduce the topic of beamformers and to collate and classify the methods.

8.2.1 Electronic Beamforming

An electronically beamforming antenna has, in general, one port for each beam. Usually these ports are well isolated. If a separate transmit or receive system is connected to each port, simultaneous independent operation in many directions can therefore be obtained. Alternatively, a single transmit or receive system can be connected to the beam ports through a multiple-way switch, giving a sequentially scanning antenna. The former configuration has an attractive property in some applications.

The creation of a multiple-beam antenna using an RF beamformer has the advantage that no devices for frequency changing are necessary. The technique therefore has the potential to be simpler and lower in cost than IF, digital or optical frequency methods. Indeed, many antenna configurations, such as lenses, have inherent multiple-beam capabilities. In these cases it is only necessary to replace the single feed by an array so that each array element forms one of the multiple beams. In other cases, such as large-array antennas for surveillance, the advantages of RF beamforming are not as clear and it is likely that optimised antenna systems may contain a mixture of RF and other beamforming methods.

The field of RF beamforming techniques encompasses two major areas:

1. Quasi-optic based with a feed array
2. Circuit based used to feed arrays

This division essentially follows that of antenna forms in continuous apertures and arrays, although the hybrid nature of many multibeam antennas represents a convergence of the two classes. Another classification is in terms of the number of Fourier transforms (FTs) that occur within the device. Any beamformer must perform an FT (in the linear space to angle sense) in distributing energy from a single feed point to the required aperture distribution. It is important to point out that, in this context, the transformation from a far-field (diffraction) pattern of the antenna to the aperture plane is described by a (angle to spatial) Fourier transform relationship. In this context the placement of feeds within a quasi-optical system requires knowledge of the wavefront transform relationships within the antenna structure. Readers are referred to specific antenna texts for a fuller discussion of these relationships.

Figure 8.8 illustrates antennas with various numbers of FTs undergone by the wavefront in its passage from the feed to the objective. As indicated, this number determines whether the feed is a phased array or simply a collection of feed points.

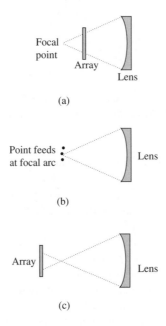

Figure 8.8 Classification of multiple-beam systems by the number of Fourier transforms: (a) zero FT with an array feed, (b) one FT with a set of point feeds and (c) two FTs with an array feed

A low-cost beam-steering system will now be presented, which is based on the principle of focal plane scanning. Figure 8.9 shows the basic concept of such a beam-steering system. Discrete lenses allow the presence of several simultaneous beams at different angles, with a simpler feed structure than phased arrays. Here, instead of using a bulky microwave lens, a planar filter lens array (FLA) as described in [22] can be used as the focusing element.

A conformal feed matrix is placed in the focal arc of a lens and fed by a feed network equipped with PIN (positive intrinsic negative) diode switches. In this way, each element of the feed matrix can be activated independently by a simple and cost-effective electronic PIN diode switch, which allows the excitation of the lens from different feed points on a grid in the focal plane.

Several constraints apply to the design of the feed arrays. For high efficiency, the lens aperture should be illuminated uniformly and a maximal flat radiation pattern in the mainlobe is required. Also, spill-over power is an important source of loss and should be minimised.

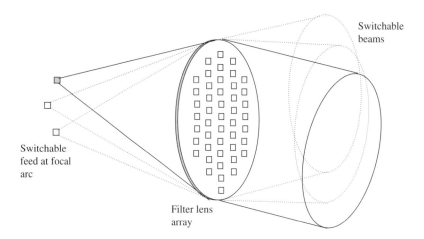

Figure 8.9   Beam-steering system with a matrix of switchable feed arrays located in the focal plane of the filter lens array. (Reproduced by permission of © 2005 IEEE [21])

Finally, the required scan resolution, limits the distance between adjacent feed arrays in the focal plane. These conditions are obviously conflicting and a compromise has to be sought. For example, a smaller beamwidth for low spill-over power can only be achieved by a larger array size at the expense of a decreased scan resolution. One way to increase the scan resolution without compromising performance is to overlap the feed arrays when grouping them into a feed matrix. As pointed out in Reference [23], overlapping can be accomplished either by interleaving parts of adjacent arrays or by sharing some of their elements [21].

Another key performance factor is scan angle capability, which is usually limited by one or more of the factors given in Table 8.1. Performance within this scan range is determined by antenna geometry, beam width, beam spacing and beam crossover level; all are important parameters. The probable system loss components are: aperture taper loss (aperture efficiency), spill-over loss, resistive loss (in both active and passive components), loss due to manufacturing errors and some additional loss related to the beam spacing, which reduces to zero for orthogonal beam sets [25]. Most practical antennas will not have perfectly orthogonal beams and will therefore incur this extra loss. It should be noted that there is another loss due to aperture phase errors which, for a reflector antenna, indicates reflector profile errors, for a lens this generally means shape errors.

Table 8.1   Factors limiting the scan angle [24]

Limitation	Cause
Array grating lobes	Insufficiently filled aperture
Array blindness	Mutual coupling or leaky-wave action
Pattern degradation (reduced gain, increased beamwidth and sidelobe level)	Reduced effective aperture size with scan, phase and amplitude errors, beamformer frequency dependence
Spill-over (in quasi-optical beamformers)	Insufficient objective or feed aperture size

As an alternative to lens beamformers, circuit beamformers use transmission lines, connecting power splitters and couplers to form multiple-beam networks. The phase shifts required to produce beam scanning are provided by lengths of transmission line.

Aperture amplitude distributions are controlled by the power splitter ratios. Two main classifications exist, namely the Blass [26] and Butler forms (see Section 8.1). The Blass matrix is far more flexible than the Butler matrix but is usually more lossy due to the presence of line terminations.

The Blass matrix consists of a number of travelling wave feed lines connected to a linear array through another set of lines, as shown in Figure 8.10. The two sets of lines are interconnected by directional couplers at their crossover points. A signal applied at a beam port will progress along the feed line to the end of termination. At each crossover point a small signal will be coupled into each element line which excites the corresponding radiating element. The path difference between the input and each element, controls the radiated beam direction. The aperture illumination is controlled by the coupling coefficients. Owing to the travelling wave nature of the network, the input match will be good and the beam set will scan with frequency.

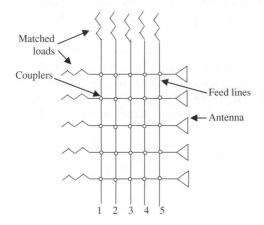

Figure 8.10   Blass matrix

Figure 8.10 shows a phase-delay device. However, a true time-delay type [27] has been described that has broader bandwidth capabilities. Beam port isolation is ensured for beam 1 because of the directivity of the couplers. However, when port 2 is excited, beam 2 is produced together with a second-order beam due to coupling through the beam 1 feed line. Blass shows that for a 75 % efficient matrix with beam separation of one beamwidth, the spurious beamlobe will be −13 dB down on the main beam and have the characteristic pointing of beam 1. If the beam spacing is increased to two beamwidths this drops to −19 dB [24].

The network design procedure consists of selecting the appropriate phase and coupling values to achieve the desired beam set. These can be simply computed for a given transfer function to achieve the minimum terminated power. However, in practice the range of coupling values available to the designer is severely limited [28].

A synthesis method [29] can be used to estimate the efficiency for arbitrary beam crossover levels. Shaped beams have also been synthesised [30]. The Blass matrix concept has been extended to form a planar, two-dimensional multiple-beam microstrip patch array [31]. In this array, microstrip patches in a resonant array formation replace the directional couplers.

Table 8.2 shows the number of hybrid couplers required by an 8 − 8 Blass matrix and planar Butler network form. Matrices have been constructed in various media, including waveguides for high power use [32, 33] and microstrips [34].

Table 8.2 Number of hybrid couplers in 8×8 matrix beamformers[24]

Matrix type	Number of hybrid couplers
Blass	64
Planar Butler	32

In addition to performing linear array scanning, the matrix can be used as a commutation device in circular and cylindrical arrays (scanning the beam around an axis) [35, 36]. Figure 8.11 shows the feed and commutating matrix arrangement. If a beam port is excited then all inputs to the commutating matrix will have uniform amplitude and an appropriate linear phase distribution. This will be transformed to a (sin X)/X distribution centred on the corresponding array port. Thus exciting consecutive beam ports results in the array distribution being scanned along the array plane of the commutating matrix, and therefore around the circular array. Table 8.3 gives some representative performance characteristics. It can be seen that there is a clear distinction between the Blass matrix and the Butler matrix. Large-aperture systems have a limited scan range whereas small antenna size allows a wide scan range.

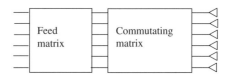

Figure 8.11 Schematic diagram of a scanning multimode array

Table 8.3 Summary of typical performance characteristics of RF beamforming systems [24]

Beamformer types	Typical scan range	Typical aperture size	Typical sidelobe level for multiple beams	Bandwidth capability
Blass matrix	$\pm 60°$	16λ	$-13\,\mathrm{dB}$	$< 1\,\%$
Butler matrix	$\pm 60°$	16λ	$-13\,\mathrm{dB}$	$> 2{:}1$

In communications, beamforming can improve the connection between a mobile device and a base station, and avoid coverage of most of the areas where no transmission is needed. Additionally, a directional antenna may be used by the receiver to improve the signal-to- interference ratio by nulling out any interference from unwanted transmitters. In both cases, only rough estimates about the size and shape of such a transformed beam can be made. Therefore, there is a need for an improved method and apparatus for beamforming in a wireless communication system. This approach offers an improved link budget, and also increased capacity which improves the possibility frequency re-use [2].

Reference [37] describes a method of transmitting information between a station with an adaptive antenna array and a receiving station. The transceiver station included an adaptive antenna array, consisting of multiple antenna elements for communication with another transceiver station. The terminal also included a controller for receiving and transmitting a digitised data stream coupled with the adaptive antenna array. The data stream included weight vector information. The arrangement also included a weight modification unit within the controller for modifying the received weight vector information, with the controller re-transmitting the modified weight vector information to another transceiver station.

One example of improving the transmission method was to use a space division multiple-access scheme. Within a transmission area of the antenna array, a spatial/angular filter was used to minimise inter- and intracell interference. In this way, a sectorised antenna array was used so that a number of fixed angular ranges were covered. However, a more sophisticated way of controlling an antenna array is by using a digital beamforming technique. Thus, the beam can be adapted to the area to be covered more flexibly and accurately.

An antenna array, usable within a base station as well as in a mobile terminal, is shown in Figure 8.12, which shows only the transmitting part of such an antenna system. The receiving part can be formed in a similar way. The antenna array here includes four antenna segments and appropriate driver circuitry. A signal sample generator receives a digital signal to be sent to a remote receiving unit. The signal sample generator generates a plurality of digital signals for each antenna element k and these digital signals are respectively multiplied at the multiplier by weight vectors. Thus, the linear combination of the data at the kth sensor can be expressed as:

$$y_n(\theta) = \sum_{k=0}^{K-1} w_k(n)x_k(n)$$

where $w_k(n)$ is the complex weight at the kth element conjugated with $x_k(n)$, which is the nth sample of the incoming signal at the kth antenna array element.

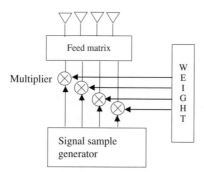

Figure 8.12   Configuration of a beamforming antenna

The processor controls the antenna to perform adaptive beam steering using multiple transmit antennas, in conjunction with receive antennas of the receiver, by iteratively performing a set of training operations. During training operations the processor causes the beamforming antenna to transmit a training sequence, while the receive antenna-array weight vector of the receiver is set and a transmitter antenna-array is switched between a set of weight vectors.

Figure 8.13 shows block diagrams of one example of both a transmitter and a receiver which are part of an adaptive beamforming multiple- antenna radio system. The transceiver includes multiple independent transmit and receive chains and performs phased array beamforming, using an array that takes an identical RF signal and shifts the phase for one or more antenna elements in the array to achieve beam steering.

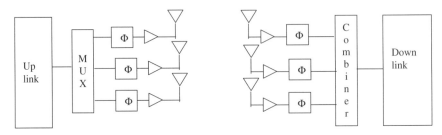

Figure 8.13   Transmitter and receiver with beamforming antennas

Each of the phase shifters produces an output that is sent to one of the power amplifiers, which amplify the signal. The amplified signals are sent to an antenna array that has multiple antenna elements. In this example the signals transmitted from the antennas are radio frequency signals between 56 and 64 GHz using quantitised phase shifters or complex multipliers.

8.3 Millimetre Wave MIMO

There are existing point-to-point wireless links that reach speeds of the order of Gb/s. For example, a 1.25 Gb/s point-to-point link using the 60 GHz band is reported in reference [38], and similar products are available in the marketplace [39]. However, MIMO (multiple input multiple output) technology can be used to increase such data rates by more than an one order of magnitude, to 10–40 Gb/s. In effect MIMO technology provides the ability for an array to support many independent communications channels as long as the elements in the array can "see" a separate link to a specific element in the communicating array. The elements in the H matrix transfer function are then (essentially) independent [40]. In addition to the natural application for communication infrastructure recovery after disasters, such wireless links offer tremendous commercial potential, as they can be used interchangeably with optical transmission equipment. For commercial applications, perhaps the greatest advantage of 10–40 Gb/s wireless links is their low cost, as they provide the bridge connections between optical links, where difficult terrain such as mountains and rivers are to be crossed or where installation costs are prohibitive, as in city centres.

When the same signal is transmitted by each antenna, it is possible to get approximately an MN-fold increase in the SNR, yielding a channel capacity equal to:

$$C \approx B \, \log_2(1 + MN \, \mathrm{SNR}_0) \tag{8.1}$$

Thus, it can be seen that the channel capacity for the MIMO system is higher than that of multiple-input single-output or single-input multiple-output. However, it should be noted here that in all four cases the relationship between the channel capacity and the SNR is logarithmic. This means that trying to increase the data rate by simply transmitting more power is extremely costly [41].

When different signals are transmitted by each antenna, it is assumed that $N \geq M$, so that all the transmitted signals can be decoded at the receiver. The critical idea in MIMO is that it is possible to send different signals using the same bandwidth and still be able to decode them correctly at the receiver. This is analogous to creating a channel for each one of the transmitters. The capacity of each of these channels is roughly equal to [40]:

$$C_{\text{single}} \approx B \, \log_2 \left(1 + \frac{N}{M} \, \text{SNR}_0 \right) \tag{8.2}$$

However, since there are M of these channels (M transmitting antennas), the total capacity of the system is:

$$C \approx MB \, \log_2 \left(1 + \frac{N}{M} \, \text{SNR}_0 \right) \tag{8.3}$$

As can be seen from Equation (8.3), a linear increase in capacity is obtained with respect to the number of transmitting antennas. Thus, the key principle here is that it is more beneficial to transmit data using many different low-powered channels than using one single high-powered channel [42]. Figure 8.14 shows the information capacity for one-input one-output, two-input two-output and three-input three-output systems. As the number of MIMO systems increases, the information capacity increases accordingly.

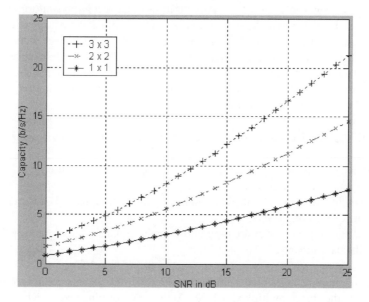

Figure 8.14 Information capacity for one-input one-output, two-input two-output and three-input three-output systems

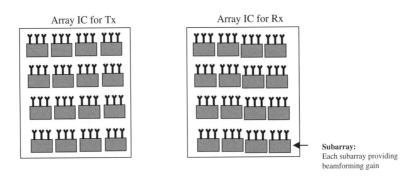

Figure 8.15   Configuration of a millimetre wave MIMO system. Each integrated circuit (IC) consists of an array of subarrays. Each subarray in a node steers a beam towards the node it is communicating with, providing beamforming gain and intersymbol interference (ISI) reduction

Figure 8.15 is an example of a millimetre wave MIMO configuration. At small wavelengths around 60 GHz, it is possible to synthesise highly directive beams with moderately sized antennas, permitting significant spatial re-use and drastically limiting multipaths. The key concepts behind this approach are as follows:

1. Adaptive beamforming. By forming a highly directive beam steerable over ten times of half-power beamwidth, the task of installation can be simplified. The directivity gains are obtained at both the transmitter and receiver by the use of adaptive antenna arrays, which are termed subarrays.
2. Spatial multiplexing. The transmitting and receiving nodes each consist of an array of subarrays, as shown in Figure 8.15. After transmit and receive beamforming using the subarrays, each subarray can be interpreted as a single virtual element in a multiple-input multiple-output (MIMO) system. As a consequence of the small wavelength, moderate separation between the subarrays ensures that each virtual transmit element sees a suitable differential response at the virtual receive array. This enables spatial multiplexing; different virtual transmit elements can send different data streams, with a spatial equaliser at the virtual receive array used to separate the streams.

Figure 8.15 shows a 4 × 4 array of subarrays at each end, with the following parameters:

- Each parallel spatial link employs quadrature amplitude modulation (QAM) with full bandwidth, transmitting at 3 Gb/s.
- A selected eight out of the 16 subarrays transmit parallel streams at 3 Gb/s, resulting in an aggregate link speed of 24 Gb/s.
- All 16 subarrays at the receiver are used in the spatial equaliser in order to separate out the eight parallel data streams.

The signal processing underlying a millimetre wave MIMO system operates from the beamforming layer to the spatial multiplexing layer. At the beamforming layer, each subarray at the transmitter synthesises a beam to point towards the receiver, and each subarray at the receiver synthesises a beam to point towards the transmitter. Once these beams have been formed, spatial multiplexing layer signal processing for the resulting virtual MIMO system can continue. Possible low-cost implementations of millimetre wave MIMO systems however, rest on ongoing advances in modern CMOS (complementary metal oxide semiconductor) technology, as well as cost-effective packaging techniques.

Beamforming and diversity using receive antenna arrays are a classical concept in communication theory, but the important role played by transmit antenna arrays, when used in conjunction with receive arrays, was pointed out by the pioneering work of Telatar [43]. Since then, three major concepts for utilising transmit antenna arrays have emerged: spatial diversity, spatial multiplexing and transmit beamforming.

Millimetrewave MIMO is different from other MIMO systems at lower frequencies in two aspects:

- Beamforming layer. The beamforming function is preferred for line-of-sight or quasi-line-of-sight channels whereas diversity is preferred for non-line-of-sight channels.
- Spatial multiplexing layer. Spatial multiplexing is obtained by focusing the receive antenna array on the different transmit antenna elements instead of relying on a rich scattering environment.

Beamforming layer signal processing for beamforming is discussed below. At millimetre waves, it is very challenging with current technology to have analogue-to-digital conversion of a signal with several GHz bandwidth at sufficient precision, for beamforming on the complex envelope. The first step is therefore to consider the architecture for the beamforming layer that combines up/down conversion with antenna phase selection.

8.3.1 Beamforming Layer

The basic building block of a millimetre wave MIMO system is a beamsteering integrated subarray. Each beamforming integrated circuit (IC) electronically steers an $M \times M$ antenna array with element spacing d, as shown in Figure 8.15, where the required M is estimated to be between 4 and 10.

The antenna directivity is proportional to its effective aperture. The effective aperture of the subarray can be increased using a telescopic dish configuration or a planar printed circuit board implementation (see Figure 8.16), while maintaining the steerability of the antenna. This provides the necessary beamforming gains to offset the higher attenuation in millimetre waves, and can be used to suppress the multipath to the possible extent.

The directivity gain of each subarray is:

$$G = \frac{4\pi A_{\text{eff}}}{\lambda^2}$$

The effective aperture A_{eff}, of the half-length spaced square array at millimetre waves is small. A_{eff} can be increased using:

(a) lenses with a large diameter (see Chapter 6) or
(b) antenna elements on a printed circuit board with a large area.

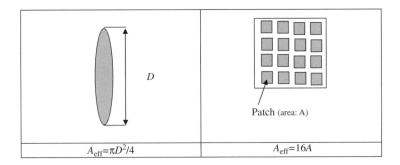

Figure 8.16   Steerable subarray configurations: (a) lens antenna and (b) planar configuration

While phased arrays at lower speeds can employ complex-valued beamforming weights at baseband, such approaches do not scale to the symbol rates and carrier frequencies of interest in this book. Therefore a row–column beam-steering IC, as depicted in Figure 8.17, is used in which two multiphase local oscillators are mixed to synthesise the millimetre wave carrier for each antenna element. Thus, the phase of the (i, j)th element of the array is given by

$$\phi(i, j) = \phi_h(i) + \phi_v(j), \quad 1 \le i, j \le M \tag{8.4}$$

where $\phi_h(i)$ is the phase for the ith row and $\phi_v(j)$ is the phase for the jth column, both chosen from a discrete set of values distributed uniformly around the unit circle. For the far-field regime, the transmit subarrays beamform towards the receiver subarrays and vice versa, which can be accomplished efficiently using a two-parameter search.

A special case of the row–column beam-steerer occurs when both the horizontal and vertical phases obey a linear profile, corresponding to steering a linear array in a specific direction; i.e. $\phi_h(i) = i\delta_h$ and $\phi_v(j) = j\delta_v$, where $\delta_h = 2\pi \sin \theta_h / \lambda$ and $\delta_v = 2\pi \sin \theta_v / \lambda$ are the phase shifts for adjacent horizontal and vertical elements, respectively, corresponding to a horizontal steering angle of θ_h and a vertical steering angle of θ_v. Here the phase increments θ_h and θ_v must also be chosen from the discrete set allowed by our hardware constraints (i.e. phase increments of $\pi/4$ or $\pi/8$, corresponding to the use of 8- and 16-phase oscillators, respectively). The minimum phase increment corresponds to the desired resolution in steering angle.

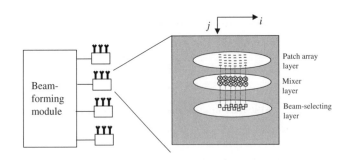

Figure 8.17   Configuration of the row–column beamsteering array [44]

8.3.2 Spatial Multiplexing Layer

The millimetre wave MIMO antenna can be designed as an array of monolithic subarrays (see Figure 8.15). Spatial multiplexing is obtained in effect by focusing the receive subarrays on to the individual transmit subarrays. Once the subarrays beam-steer along the desired direction, they can be considered to be antenna elements of a virtual MIMO system. An $N \times N$ array of subarrays with lateral spacing D has dimensions $(N-1)D \times (N-1)D$. To realise the desired spatial multiplexing, each of the N virtual transmit elements must see a different N receive array response, in order to be able to separate out the different transmitted streams. The Rayleigh criterion in imaging [45] determines the minimum spacing between transmit elements so that they can be resolved by the receive array with no coupling effect. In the case of sub-Rayleigh spacing, the correlation between the responses at the receiver for two different (virtual) transmit elements can be derived. For uniform linear arrays (ULA) aligned to the broadside of each other, as displayed in Figure 8.18, spatial angular separation of two transmitters is:

$$\delta\theta = D/R$$

Then, the signal phase separation at the receivers is

$$\delta\phi_e = \delta\theta \times 2\pi D/\lambda$$

If the phase difference at the receiver is $\delta\phi_e = \pi$ (e.g. $D = \sqrt{\lambda R/2}$), then simply in-phase combining the receiver signals to point the receiver array at the desired transmitter will result in (ideally) 100% suppression of the signal from an undesired transmitter. This corresponds to the Rayleigh criterion in diffraction-limited imaging.

The above circuit functions are integrated in an array format (see the block diagram in Figure 8.17) to support an $M \times M$ antenna matrix. Consequently, the highly regular floor plan of the top-layer layout of the beam-steering IC eases the chip-to-board interface design and thus accommodates the matching networks for the antenna matrix. This high layer of parallelism and complexity can be implemented using modern CMOS technology. The main advantage of using CMOS for millimetre wave systems up to 100 GHz (see Section 9.3) is its capability to

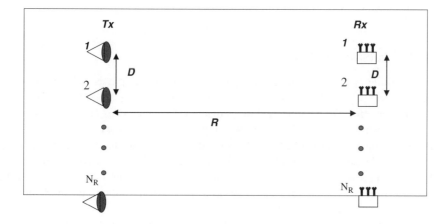

Figure 8.18 Geometry of the linear array MIMO system at the spatial multiplexing layer

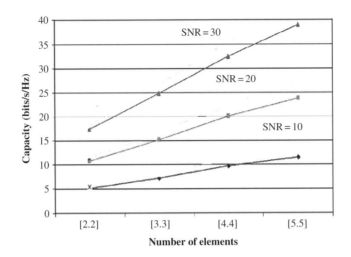

Figure 8.19 Capacity of the single point-to-point link for different modulation schemes, selecting two, three, four and five elements at Tx and Rx. (Reproduced by permission of © 2007 IEEE [47])

integrate massively parallel transceiver arrays for directivity gain and adaptive beamforming. The beam-steering ICs are already in development using a 90 nm CMOS technology [46]. Preliminary system level simulation shows a 90 dB gain with a 32×32 overall array -4×4 beam-steering ICs with each one supporting an 8×8 antenna matrix [44].

Figure 8.19 shows the median value of the capacity of the measured channel based on the number of elements per array ($[Tx \cdot Rx] = [2 \cdot 2], [3 \cdot 3], [4 \cdot 4]$ and $[5 \cdot 5]$), with three different SNR values: 10, 20 and 30 dB. One can see that the capacity of the measured channel is slightly lower than that of the MIMO Rayleigh channel. The difference between the experimental results and the theoretical values increases as the SNR increases. This difference is mainly due to the fact that the distance between the elements is not infinite but is equal to one wavelength, that the angles of arrival and departure of the signals are not uniformly distributed and that the channel is not ideal (a finite number of multipaths).

8.4 Mode-Tracking Antennas

High-speed mobile applications are required to be spectrum and power efficient in order to have accurate pointing of the antenna from base stations or access points towards mobile devices. Thus some form of antenna tracking is desired. To achieve this goal, there is a variety of microwave sensing techniques in use that are employed to detect and correct pointing errors. These include conical scan, peripheral feed horns, step track and the more expensive and more accurate multimode monopulse systems. Depending on the application, all of these approaches suffer from certain disadvantages. These limitations include pointing inaccuracy, a relatively slow response time, the deleterious effect of atmospheric scintillation and implementation complexity in the case of low-cost systems.

A special technique for RF beam scanning is described [48], which takes the form of generating proportions of higher-order waveguide modes within the antenna feed horn in order to electronically squint the secondary pattern in the azimuth and elevation planes. This technique involves a high-performance horn configuration, which incorporates the sequential lobing or beam-shift tracking capability of multimode feeds to provide an accurate, yet fast, response method of acquisition and fine pointing of a base station antenna. As with other methods of tracking, gross changes of antenna pointing are undertaken by incremental use of the main axis drive motors. A mode generator is introduced within the primary-feed system to produce the desired higher-order mode selectively.

The principal advantages of this form of tracking are:

- There is no need for a separate and expensive tracking receiver.
- Relatively simple microwave components are employed in the feed chain.
- All solid-state technology using PIN diodes, as can be seen in Section 4.2 on a multimode horn antenna, are controlled by a microprocessor that is compatible with existing control systems.
- For fast electronic acquisition, use of PIN diodes leads to the capability for extremely high switching speeds and implies that the system can cope with all effects encountered in the tracking movement, thus maintaining equivalent isotropically radiated power (EIRP) stability.
- Greater stepping speed provides larger sample gathering in a short time and hence improves the noise performance of the system.
- Greater ability to optimise antenna pointing is provided over very short time intervals with consideration of real-time tracking, e.g. during wind gusts or rapid platform movement in the case of oil rig, ship-borne or vehicle-mounted antennas.
- Less volume is required, thus easing accommodation and mounting aspects.
- The feed system can be easily retrofitted to an existing antenna.
- A pointing accuracy results, which may approach, or equal, that of the traditional monopulse system and be superior to existing conical scan or step track systems.
- For the end user, employment of this method of electronic tracking leads to both a reduction in the required motion during acquisition, less demand and wear on motor drives, rack, etc., thus minimising the maintenance times and replacement of worn parts.
- A better fault tolerance is achieved. In the unlikely event of a single diode failure condition in either, or both, the azimuth and elevation planes, a mobile tracking device is still maintained.
- Depending on the antenna configuration, replacement diode assemblies may be fitted while the antenna is operational.

For the principle of operation of an overall system, the operation of the beam squint tracking system may firstly be explained with reference to Figure 8.20, which shows the locations of the beam relative to the true antenna boresight in contour plot form. The central axis (0) of the circle represents the direction of the boresight, and positions away from the axis represent angles from the boresight. The antenna feed includes four sequentially operated PIN diode-controlled mode generators, which squint the beam to four different angles, 1, 2, 3 and 4. It should be appreciated that the axial directions 1, 2, 3 and 4, indicated in Figure 8.20, are associated with the maxima of reception; a signal located away from the boresight axis is still received, but the level is reduced marginally due to its displacement (the effect of induced mode conversion from the mode generator is to translate the antenna's secondary pattern through an angle of $0.06°$, corresponding to a beam shift of 20 % of the half-power beamwidth).

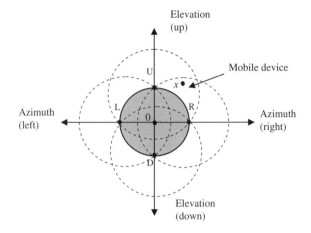

Figure 8.20 Polar diagram showing the directional locations of the squinted secondary beam peak levels (U, D, R, L) relative to the boresight (0) and position of the incoming mobile terminal (x)

Consider a beacon signal transmitted from a mobile device that may be located at an off-boresight position x in Figure 8.20. It is assumed that the position of the incoming beacon is not known at the receiving antenna. To locate the mobile device position, each one of four higher-order waveguide mode generators in the base station begins a search pattern in which the reception direction of each beacon signal strength is switched from the true boresight (0) to each of the positions U, D, R and L in turn [48]:

U = elevation up
D = elevation down
R = azimuth right
L = azimuth left

The intensity of the beacon signal at each beam position is detected by the communications receiver, sampled by an analogue-to-digital (A/D) converter and each measurement is passed to a microprocessor where it is stored in conjunction with its coordinate direction. The system block diagram portraying the arrangement of the electronically controlled PIN diode waveguide mode generators in the feed chain, the communications receiver, the analogue-to-digital converter, the microprocessor and the steering control mechanism is shown in Figure 8.21. The rapid switch-and-measure sequence enables the whole search pattern to be completed in a small fraction of a second. Although the mobile device is in general always moving, no substantial change in position occurs with in this time frame. Thus, the four measurements of the search pattern (U, D, R and L) can be regarded as simultaneous.

From the polar diagram (Figure 8.20) it is evident that the beam squinted positions, directions U (elevation/up) and R (azimuth/right), will give stronger signals than in the corresponding directions D (elevation/down) and L (azimuth/left). Furthermore, beam position U will produce a stronger signal than the direction of R. Using the data accumulated concerning the off-axis performance of each direction during a single frame, the coordinate position x is computed

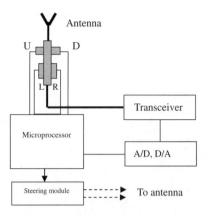

Figure 8.21 Electronic tracking system simplified block diagram for receive channels [49]

and this provides an error signal for the feedback loop operating the steering. To establish the position estimates, if a parabolic main beam shape is assumed, it can be shown that the tracking error is [49]

$$\alpha = \frac{\Delta P}{k\theta} \quad \text{(volts/degree)}$$

where ΔP is the difference in signal level measured at the two squinted positions in a common plane, θ is the beam deflection and k is a constant that depends on the antenna dimensions and operating frequency.

As each PIN diode mode generator operates rapidly, this makes it possible to obtain a sequence of positions at short time intervals which can also provide the necessary data for a prediction algorithm [50] (if this beam squint system is backed up by a complementary intelligent smooth step track system). In the case of an access point using well-established information about radiation distribution and reflection characteristics inside a room, the algorithm can predict the direction of the optimum wireless link. Furthermore, it is also possible to estimate the time required for a steering operation and hence to obtain a predicted final position where the mobile device will be at the end of the steering operation. The predicted position constitutes a suitable input for the feedback loop. The advantage of the beam squint approach is clearly evident as all directional data concerning the mobile device are obtained using electronic methods.

This reduces the use of the motor drives and obtains more data in a shorter time, whereby the performance of pointing prediction algorithms is enhanced. It simplifies searching during steering, since fundamentally different systems are used for the two operations.

A feed network is required (outlined in Chapter 4) that produces the desired beam shift necessary to define the magnitude and direction of an antenna's pointing error, based on prescribed and equal deviations in the azimuth plane (Δx) and in the elevation plane (Δy). In a practical application, such a feed (Figure 8.22) would also handle a receive and transmit link, as well as the embedded beacon or tracking signal. Commencing with the conical horn, this feed comprises the serial connection of a taper, a higher-order mode generator or converter and a conventional orthomode transducer (OMT).

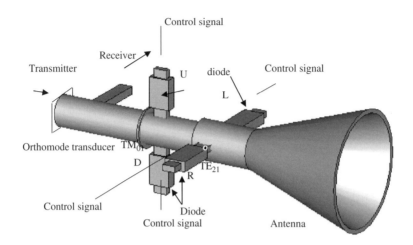

Figure 8.22 Example of an RF feed chain incorporating an electronic beam shift tracking for vertically polarized signals in a millimetre wave access point

To simplify the understanding of the action of the mode generator, it is best to view the device operating in a transmit sense. In this case, a fundamental mode is injected into the circular waveguide of the OMT. The next component towards the radiating horn is a mode generating section. For the purpose of this discussion, the mode generator may be considered as a "black-box" into which the fundamental mode signal is injected.

The output from the "black-box" will be a proportion of the original fundamental mode together with a component of a desired higher-order mode. This derived higher- order mode, in combination with the fundamental mode, propagates to the horn and produces a modified aperture illumination. Assuming the higher-order mode is present at the horn aperture in the correct amplitude and quadrature phase relationship relative to the fundamental mode, then the modified aperture illumination can be arranged to produce an overall phase tilt.

To obtain sufficient tracking information to define the antenna pointing error fully, it will be necessary to produce and analyse beam shifts in two orthogonal directions. For tracking in the other orthogonal cardinal plane, a similar arrangement of mode generators is used.

Since reciprocity will apply when the antenna is operating in the receive sense, the power level variation that occurs with the beam shift can then be detected by the conventional communications receiver, as indicted in Figure 8.21. In operation, the mode generator is controlled electronically in a sequential manner, to switch the modes. The mode generator section comprises a central circular waveguide with a number of short-circuited rectangular waveguide auxiliaries coupled at the periphery of the circular waveguide in selected positions. Each auxiliary waveguide connected to the central circular waveguide is terminated by a PIN diode. The mode generator is activated by reversing the bias of the PIN diode and the incoming higher-order mode (from the horn) is converted into the fundamental mode by introducing an asymmetry to the circular waveguide.

Two discrete mode generator sections are used to achieve orthogonal plane ($\Delta x/\Delta y$) beam deviations in linear polarisation. Connected directly to the horn throat is the TE_{21} mode generator in which one pair of diametrically opposed auxiliary rectangular waveguides (aligned with the horizontal plane) are coupled "longitudinally" to the periphery of the central circular waveguide. The TM_{01} mode generator has a second pair of auxiliary arms (transversely coupled), which are again diametrically opposed to each other. In this arrangement, the auxiliary waveguides in the TM_{01} mode generator provide elevation plane (Δy, up and down, respectively) tracking information, and the lateral auxiliary waveguides of the TE_{21} mode section provide azimuth plane (Δx, left and right) beam deviation. In operation, PIN diodes in each auxiliary arm operate in turn while the other three remain inoperative.

The beacon receiver output then varies in synchronisation with the switching of each auxiliary waveguide so that tracking information is gathered (the time multiplexed frame data rate, which is dependent upon particular system requirements, may be varied from a low rate to many millions of samples per second).

The complete feed system incorporating TM_{01} and TE_{21} mode generators was characterised at the beacon frequency for the usual principal H, E and diagonal plane radiation patterns. Each mode generator is rendered inactive, i.e. all diodes are in the "off" state in order to establish that their presence has no effect on propagation in the central circular waveguide. While the inherent diagonal plane peak cross polarisation is better than $-40\,dB$, the quality of balance between each pair of mode generator arms is also good, exhibiting a peak value of $-35\,dB$ in the principal plane.

Dealing with each active diode pair in turn, the lengths of the terminating rectangular waveguide short-circuits are individually optimised to establish the desired level of mode conversion. This mode conversion is measured in the far field as induced cross polarisation in the principal planes. A photo of a mode tracking antenna is shown in Figure 8.23.

ERA's 20/30 GHz electronic beamsquint tracking feed for EURECA.

Figure 8.23   Mode tracking antenna. (Reproduced by permission of © 2007 ERA Technology Ltd, from a research project at ERA Technology Ltd, http://www.era.co.uk/)

Table 8.4 Summary of various classes of smart antennas

Spatial technique	Aspects	Pros	Cons
Beam switching	Switching among set of predefined fixed multiple beams	Easy to implement. No phase shifter is needed. Stable performance when all users have low/ similar data rates	Underperforms in systems with multirate services where both low data rate speech service and high data rate applications are simultaneously supported since beams cannot track users
Beam steering	Main beam is directed towards the desired user	Weight is the same as the steering vector Provides array gain and steering gain for spatially white interference environments	Requires direction of arrival estimation of the desired user Suboptimal performance in terms of SNR or SIR
MIMO	The most adaptive scheme SNR and SINR are optimised based on some given criterion	Increase diversity and capacity Optimal weight vector and optimal performance	Computationally intensive. Multiple transceivers are needed

SINR: signal to interference plus noise ratio.

In summary, this chapter concludes by comparing the various classes, both in the light of currently obtained results and with respect to the underlying trends (Table 8.4).

References

[1] B. Neekzad, K. Sazrafian-Pour and J. D. Baras, 'Clustering Characteristics of Millimeter Wave Indoor Channels', IEEE Wireless Communications and Networking Conference, 2008.

[2] M. A. Beach, A. J. Copping, D. J. Edwards and K. W. Yates, 'An Adaptive Antenna for Multiple Signal Sources', 5th International Conference on *'Antennas and Propagation'*, York, England, 30 March–2 April 1987, pp. 347–350.

[3] A. M. Street, A. P. Jenkins, J. Thornton and D. J. Edwards, 'Low Cost Adaptive Antenna Systems for Indoor Mobile Wireless Communications', AP2000 Millennium Conference on *'Antennas and Propagation'*, Davos, Switzerland, 9–14 April 2000, ESA.

[4] http://dept106.eng.ox.ac.uk/wb/pages/research/microwave/radar-and-antennas/adaptive-antennas.php

[5] J. Butler and R. Howe, 'Beamforming Matrix Simplifies Design of Electronically Scanned Antennas', *Electronics Design*, **9**, 1961, 170–173.

[6] J. P. Shelton and K. S. Kelleher, 'Multiple Beams for Linear Arrays', *IRE Transactions on Antennas and Propagation*, March 1961, 154–161.

[7] P. J. Muenzer, 'Properties of Linear Phased Arrays Using Butler Matrices', *Standard Elektrik*, A. G. Lorenz, Stuttgart, NTZ, **9**, 1972, 419–422.

[8] T. MacNamara, 'Simplified Design Procedure for Butler Matrices Incorporating 90" or 180" hybrids', *IEE Proceedings H, Microwaves, Antennas and Propagation*, **134**(I), 1987, 50–54.

[9] J. Shelton and J. Hsiao, 'Reflective Butler Matrix', *IEEE Transactions on Antennas and Propagation*, **AP-27**(5), 1979, 651–659.

[10] J. R. F. Guy, 'Proposal to Use Reflected Signals through a Single Butler Matrix to Produce Multiple Beams from a Circular Array Antenna', *Electronics Letters*, **28**(5), 1985, 209–211.

[11] S. Ohshima, Y. Asano, T. Harada, N. Yamada, M. Usui, H. Hayashi, T. Watanabe and H. Iizuka, 'Phase-Comparison Monopulse Radar with Switched Transmit Beams for Automotive Application', *IEEE MTT-S International Microwave Symposium Digest*, **4**, 1999, 1493–1496.

[12] G. N. Hulderman, 'Stepped Beam Active Array Antenna and Radar System Employing Same, US Patent 5583511, 1996.

[13] R. Roy and T. Kailath, 'ESPRIT – Estimation of Signal Parameters via Rotational Invariance Techniques', *IEEE Transactions on Acoustics and Speech Signal Process*, **37**, 1989, 984–995.

[14] Y. Asano, S. Ohshima, T. Harada, M. Ogawa and K. Nishikawa, 'Proposal of Millimeter-Wave Holographic Radar with Antenna Switching', *IEEE MTT-S International Microwave Symposium Digest*, **2**, 2001, 1111–1114.

[15] N. Kikuma, *'Adaptive Signal Processing with Array Antenna'*, Science and Technology Publishing Company, 1998.

[16] S. Ohshima, Y. Asano and K. Nishikawa, 'A Method for Accomplishing Accurate RCS Image', *IEICE Transactions on Communications*, **E79-B**, 1996, 1799–1805.

[17] N. Yamada, Y. Asano, S. Ohshima and K. Nishikawa, '3-Dimensional High-Resolution Measurement of Radar Cross Section for Car in 76 GHz Band', Proceedings of the 2003 IEICE General Conference, B-1-11, 2003.

[18] Masaru Ogawa, Yoshikazu Asano, Shigeki Ohshima, Tomohisa Harada and Naoyuki Yamada, 'Electrically Scanned Millimeter-Wave Radar with Antenna Switching', *Electronics and Communications in Japan*, Part 3, **89**(1), 2006, Translated from Denshi Joho Tsushin, *'Gakkai Ronbunshi'*, Vol. J88-A, No. 2, February 2005, pp. 237–246.

[19] Graham Brooker, Mark Bishop and Steve Seheding, 'Millimetre Waves for Robotics', Proceedings of 2001 Australian Conference on *'Tobotics and Automation'*, November 2001, pp. 91–97.

[20] K. Yamane, A. Sanada and K. Ohkubo, 'A Holographic Imaging Method for Automotive Radar', *Transactions of IEICE*, **J81-B-II**, 1998, 805–813.

[21] C. Barth, K. Caekenberghe and K. Sarabandi, 'A Novel Low-Cost Millimeter-Wave Beam-Steering System', IEEE.APS International Symposium, Vol. 4B, 2005, pp. 31–34.

[22] A. Abbaspour-Tamijani, K. Sarabandi and G. M. Rebeiz, 'A Planar Filter–Lens Array for Milimeter-Wave Applications', *IEEE Transactions on Antennas and Propagation*, **1**, 2004, 675–678.

[23] A. Abbaspour-Tamijani and K. Sarabandi, 'An Affordable Millimeter-Wave Beam-Steerable Antenna Using Interleaved Planar Subarrays', *IEEE Transactions on Antennas and Propagation*, **51**(9), September 2003, 2193–2202.

[24] P.S. Hall and S.J. Vetterlein, 'Review of Radio Frequency Beamforming Techniques for Scanned and Multiple Beam Antennas', *IEE Proceedings, Part H*, **137**(5), October 1990, 293–303.

[25] W. D. White, 'Pattern Limitation in Multiple Beam Antennas', *IRE Transactions on Antennas and Propagation*, July 1962, 430–436.

[26] J. Blass, 'Multi-directional Antenna – New Approach Top Stacked Beams', *IRE International Convention Record*, Part 1, 1960, 48–50

[27] R. C. Hansen, *'Microwave Scanning Antennas'*, Vol.111, Academic Press, New York, 1966, p. 246.

[28] M. Fassett, L. J. Kaplan and J. H. Pozgay, 'Optimal Synthesis of Ladder Network Array Antenna Feed Systems', APS Symposium, Amherst, Massachusetts, 11–15 October 1976, pp. 58–61.

[29] N. Inagaki, 'Synthesis on Beam Forming Networks for Multiple Beam Array Antennas with Maximum Feed Efficiency', IEE International Conference on *'Antennas and Propagation'*, ICAP 87, March 1987, pp. 375–378

[30] P. J. Wood, 'An Efficient Matrix Feed for an Array Generating Overlapped Beams', IEE International Conference on *'Antennas and Propagation'*, ICAP 87, March 1987, pp. 371–374.

[31] S. J. Vetterlein and P. S. Hall, 'Novel Multiple Beam Microstrip Patch Array with Integrated Beamformer', *Electronics Letters*, **25**(17), 1989.

[32] P. E. K. Chow and D. E. N. Davis, 'Wide Bandwidth Butler Matrix Network', *Electronics Letters*, **3**, 1967, 252–253.

[33] R. Levy, 'A High Power X Band Butler Matrix', *Microwave Journal*, April 1984, 135.

[34] J. R. Wallington, 'Analysis, Design and Performance of a Microstrip Butler Matrix', European Microwave Conference, Brussels, 4–7 September 1973, Vol. 1, pp. A1431–A.1434.

[35] G. Skahill and W. D. White, 'A New Technique for Feeding a Circular Array', *IEEE Transactions on Antennas and Propagation*, **AP-23**, March 1975, 253–256.

[36] B. Sheleg, 'A Matrix Fed Circular Array for Continuous Scanning'. IEEE APS Symposium, Boston, Massachusetts, 9–11 September 1986, pp. 7–16.

[37] Antoine J. Rouphael, 'Adaptive Beamforming in a Wireless Communication system', US Patent 20040204103.

[38] K. Ohata, K. Maruhashi, M. Ito, S. Kishimoto K. Ikuina, T. Hashiguchi, K. Ikeda and N. Takahashi, '1.25 Gb/s Wireless Gigabit Ethernet Link at 60 GHz-Band', *IEEE MTT-S International Microwave Symposium Digest*, **1**, June 2003, 373–376.

[39] Proxim-Wireless™, 'Gigalink Series – Alternative to Fiber up to Gigabit Speeds', DS 0806 GIGALINK USHR.pdf, 2006.

[40] John G. Proakis, *'Digital Communications'*, McGraw-Hill, New York, 4th edition, 2000.

[41] Angel Lozano, Farrokh R. Farrokhi and Reinaldo A. Valenzuela, 'Lifting the Limits on High-Speed Wireless Data Access Using Antenna Arrays', *IEEE Communications Magazine*, September 2001, 156–162.

[42] Gregory D. Durgin, *'Space–Time Wireless Channels'*, Prentice-Hall, New Jersey, 2003.

[43] E. Telatar, 'Capacity of Multi-antenna Gaussian Channels', Technical Report, AT&T Bell Labs, 1995.

[44] Eric Torkildson, Bharath Ananthasubramaniam, Upamanyu Madhow and Mark Rodwell, 'Millimeter-Wave MIMO: Wireless Links at Optical Speeds' (Invited Paper), Proceedings of the 44th Annual Allerton Conference on *'Communication, Control and Computing'*, Monticello, Illinois, September 2006.

[45] J. D. Kraus, *'Radio Astronomy'*, 2nd edition, Cygnus-Quasar, 1986, pp. 6–19.

[46] C. Carta, M. Seo and M. Rodwell, 'A Mixed Signal Row/Column Architecture for Very Large Monolithic mm-Wave Phased Arrays', IEEE Lester Eastman Conference on *'High Performance Devices'*, August 2006.

[47] Sylvain Ranvier, Jarmo Kivinen and Pertti Vainikainen, 'Millimeter-Wave MIMO Radio Channel Sounder', *IEEE Transactions on Instrumentation and Measurement*, **56**(3), June 2007, 1018–1024.

[48] D. J. Edwards and B. K. Watson, 'Electronic Tracking Systems for Microwave Antennas', British Patent Application 8414963, 12 June 1984.

[49] R. Dang, B. Watson, I. Davis and D. J. Edwards, 'Electronic Tracking System for Satellite Ground Stations', European Microwave Conference, 1985, pp. 681–687.

[50] P. M. Terrell and D. J. Edwards, 'The Smoothed Step-Track Antenna Controller', *International Journal of Satellite Communications*, **1**(2), 1983, 133–139.

9

Advanced Antenna Materials

At millimetre wave frequencies, materials and integration techniques in RF systems are subject to more demanding performance constraints. One example in printed technologies is substrate water absorption, which above 10 GHz can cause unacceptable losses in elements such as antennas, filters and transmission lines, particularly over extended periods of time and under conditions of varying humidity. Many materials, which are used for 2.4 and 5.8 GHz wireless local area networks, may also have high dielectric loss for millimetre wave applications. In addition, the market demands improved performance at low cost. Thus, new material technologies must be identified that can simultaneously fulfil these requirements of performance, frequency and environmental invariance, and low cost.

Millimetre wave systems tend to be designed around two major philosophies: system-on-chip (SoC) and system-on-package (SoP). System-on-chip is a fully integrated design approach with RF passive components and digital and/or optical functions on-wafer [1]. System-on-package incorporates analogue components into a multilayer dielectric material and integrates chips within or on the same dielectric packaging material [2]. For the system-on-chip approach, especially for millimetre waves, gallium arsenide (GaAs) is normally used owing to the high cut-off frequency performance it offers digital transistors, and for the lower substrate loss it provides for analogue components. However, GaAs is much more expensive than silicon, and using large areas of the substrate for analogue components is not deemed to be cost effective. Silicon germanium (SiGe) on either CMOS/BiCMOS-grade silicon (Si) or high-resistivity Si is a lower cost option than GaAs for some applications, but it is still a relatively lossy substrate for passive RF components. System-on-package (SoP) modules solve the major shortfalls of system-on-chip (SoC), by providing a low-loss substrate material for the RF passives and a unique space-saving capability for chip integration in or on the substrate.

The industry standard for circuit boards, i.e. FR4, becomes unacceptable due to prohibitively large losses and is not suitable for SoP in the high gigahertz range. Low temperature co-fired ceramic (LTCC) has attractive electrical characteristics, can support dense multilayer circuit integration and has very good package hermeticity, but the cost is also relatively high [3]. Notwithstanding the cost issues, it is a major candidate in millimetre wave system integration, but owing to its limitation in terms of design rules (i.e. minimum line width and spacing)

Millimetre Wave Antennas for Gigabit Wireless Communications   Kao-Cheng Huang and David J. Edwards
© 2008 John Wiley & Sons, Ltd

compared to thin-film technologies, the design at millimetre wave frequencies becomes challenging. Nevertheless, the very mature multilayer construction capabilities of LTCC enable the replacement of broadside coupling mechanisms by vertical coupling, and make LTCC a competitive solution to meet millimetre wave design requirements.

The recent introduction of liquid crystal polymer (LCP) substrates and packaging materials that have low loss (at least tan δ < 0.005) and multilayer construction capabilities, has meant that they can be considered for vertical integration, and they have the added benefit of low water absorption properties and good mechanical properties. LCP technology combines the properties of polymers with those of liquid crystals. These hybrids show the same mesophase characteristics as ordinary liquid crystals, while still retaining the versatile properties of polymers.

As a final consideration, the development of new millimetre wave applications in CMOS, metamaterials and high-temperature superconductors will also be described in this chapter.

9.1 Low-Temperature Co-fired Ceramics

LTCC as a ceramic multilayer technology has great potential for micro- and millimetre wave applications. In spite of being a very mature technology, LTCC has recently gone through large improvements in material development and has become widely available for communication equipment manufacturers through LTCC foundries. The competitive price of materials and production costs makes LTCC an attractive solution for system-in-package (SiP) and multichip modules (MCM) [4]. LTCC circuits can consist of 2 to 100 layers. LTCC substrates are robust, hermetic and environmentally stable. These features and other favourable characteristics are utilised to develop compact and efficient modules for communication and sensor applications.

Modern millimetre wave communications require a low manufacturing cost, excellent performance and a high level of integration [5]. LTCC system-on-package (SoP) [6] enables these requirements to be met. The approach offers great potential for passive integration and enables microwave devices to be fabricated with high reliability and reproducibility, while maintaining a relatively low cost base. Numerous publications [6, 7] have dealt with the development of three-dimensional LTCC passive components that are the critical building blocks in multilayer high-density architectures.

It is clear that millimetre wave antenna and circuits need to have high accuracy and have small tolerance for implementation. Therefore, it is worth understanding the whole processing history of LTCC as this could be useful for device debugging and optimisation. The whole process can be divided into the following steps (see Figure 9.1).

1. Substrate Preparation

The green sheets (unprocessed material) are usually shipped on a roll; the tape has to be unrolled on to a clean, stainless steel table. The sheet is then cut with a sharp blade, laser or punch into parts (these parts have to be a little larger than the blank size, if the material needs to be pre-conditioned). If a laser is used, its power should not be too high as it could cause damage to the sheets. Some tapes need to be pre-conditioned, such as baking the raw material for about 30 minutes at 120 °C (subject to the particular material). Normally the tapes

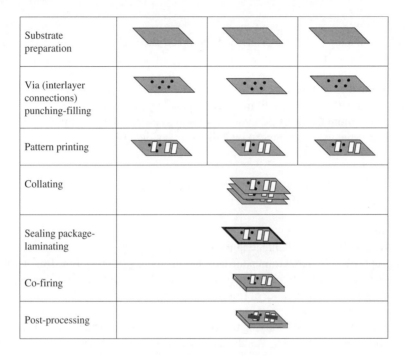

Substrate preparation			
Via (interlayer connections) punching-filling			
Pattern printing			
Collating			
Sealing package-laminating			
Co-firing			
Post-processing			

Figure 9.1 LTCC processing

are shipped with an applied foil/bake sheet, which has to be removed before lamination; some processors use this foil as a filling mask for the vias. In the meantime, a blanking die is used to create orientation marks and lamination tooling holes (and the final working dimension in case of tapes that are to be pre-conditioned). Arbitrary shapes of cavities and windows can be formed with the laser at this stage.

2. Via Forming

Vias may be punched or drilled with a low-power laser. They can then be filled with a conventional thick-film screen printer or an extrusion via filler.

- In the first case the tape has to be placed on a sheet of paper which is placed on a porous stone; a vacuum pump then holds the tape in place and is used as an aid for filling. In this approach, the vias must have a larger diameter than the tape thickness. The smallest possible size of vias to be filled also depends on the viscosity of the filling paste.
- The second method of filling the vias is to use a special material (an extrusion via filler) that works with pressures of about 4 to 4.5 bar.

Both methods need to have a stainless steel mask.

For the filling of blind vias (i.e. vias that do not go through substrates), it is advisable to form the holes in the masks a little smaller than the diameter of the blind vias. Otherwise problems could occur with the filling rate.

3. Printing

Co-fireable conductors, etc., are printed on the green sheet using a conventional thick film screen printer. As with the via printing process, a porous stone is used to hold the tape in place. Printing of the conductor can have high resolution because of the flatness and solvent absorption properties of the tape. After printing, the vias and conductors have to be dried in an oven at 80 to 120 °C for 5 to 30 minutes (depending on the material); some pastes need to level (flow) at room temperature for a few minutes before drying. In this process, resistors may vary their value when terminated with different conductors.

4. Collating

Each layer constructed in this way is placed in turn over tooling pins. Some foundries use heat to fix the sheets one on top of the other.

5. Lamination

There are two possibilities in laminating the tapes [8].

The first approach is named uniaxial lamination; the tapes are pressed between heated platens at 70 °C, 200 bar for 10 minutes (typical values). This method causes higher shrinking tolerances than the second method, of isostatic lamination (see below). The main challenge in this process is the flowing of the tape, which results in high shrinkage (especially at the edge of the wafer) during the firing, and varying thicknesses of each layer (which causes hard problems on the high-frequency sector). Thus uniaxial lamination can possibly cause dimension changes with cavities/windows.

The second approach is to use an isostatic press. In this approach the stacked tapes are vacuum packaged in a foil and pressed in hot water (similar temperatures and times are used as in the uniaxial press). The pressure is about 350 bar.

It should be noted that deep cavities and windows need to have an inlay during the lamination process.

6. Co-firing

Laminates are fired in one step on a smooth, flat setter tile. The firing should follow a specific firing profile (which means that the temperature must be controlled as a function of time and varied according to the details of the process). This means that a programmable box kiln must be used.

A typical profile shows a (slow) rising temperature (about 2–5 °C per minute) up to about 450 °C with a dwell time of about one to two hours, during which the organic burnout (binder) takes place; then the temperature has to be raised to 850–875 °C with a dwell time of about 10 to 15 minutes. The whole firing cycle lasts between three and eight hours (depending on the material; large/thick parts require a modification of the firing profile).

Note that resistor pastes need to have especially well defined firing conditions (temperatures), otherwise the final resistance value varies enormously.

7. Post-processing

Some materials need to be post-fired; this means the paste is to be applied after firing the tape and then a second firing is undergone. Post-firing thick-film resistors are normally used in 90 ° hybrid couplers and smart antennas. Resistor pastes need to have defined firing conditions (temperatures) to achieve the correct resistance. Solder conductors are also processed at the post-firing phase.

If the fired parts have to be cut into smaller pieces or other shapes, there are three different ways to achieve this:

1. The first is to use a post-firing dicing saw, which is a common method and works very well for rectangular shapes; it holds tight outside dimensional tolerances and allows high-quality sharp edges.
2. The second possibility is to use an ultrasonic cutter; this approach generally achieves low tolerances although the technique allows for the production of unusual shapes. This process is very slow and expensive.
3. The third method uses a laser to cut the fired tape; while the tolerances are tight, the quality of the edges is not good.

At this point some particular features of LTCC will be discussed, including cavities, crosshatched ground planes and vias.

Cavity design is one of the attractive features for LTCC and a typical example will be considered, as shown in Figure 9.2. Electronic components can be integrated within cavities for both functional and environmental purposes. As a general rule the cavity corner is normally rounded. It is usually recommended for the corner to have a radius larger than $100\,\mu$m. An embedded air cavity [9] and a staggered via structure can be adopted for a reduction in shunt capacitance and discontinuity. However, this is complicated by concerns of deformation during lamination and firing, particularly for thin or deep walls (the lamination and firing processes produce distortions that can alter the behaviour and the integrity of tall thin-walled structures).

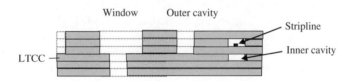

Figure 9.2   Six-layer LTCC with cavities and windows

Crosshatched ground planes (i.e. discontinuous or perforated ground planes) are an attractive approach for LTCC as they use less precious metal and, most importantly, improve the robustness of the ceramic bonding and reduce distortion in the process. It is typical that the lines of the crosshatching (the conductor) are half as wide as the spaces, and therefore there is only 55 % metal coverage of the ground area. The savings in metallisation are therefore significant. Solid areas are, however, required for electromagnetic reasons at via landings and are added below a patch or other antenna elements, as shown in Figure 9.3.

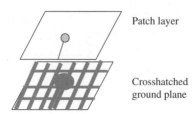

Figure 9.3   LTCC gridded or crosshatched ground plane showing continuous ground planes at features

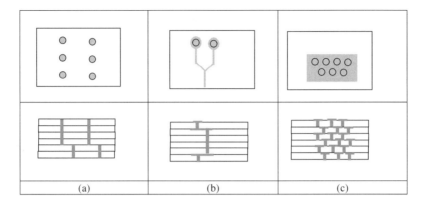

Figure 9.4 (a) RF shield vias, (b) signal vias and (c) thermal vias. The upper row shows the top view and the lower row is the cut view

There are three types of vias, as shown in Figure 9.4:

1. RF shield vias. Such vias are used to isolate radiation coupling. For example, when a milli-metre wave patch antenna is fabricated on a larger-size LTCC substrate, its radiation pattern may be significantly affected by the diffraction of surface waves at the edge of the finite ground substrate [10]. Also, the excitation of strong surface waves could cause unwanted coupling between the antenna and other nearby components on the circuit board, thus degrading the performance of the integrated module as a whole [11]. In this case, RF shield vias can be used to connect a hollow square of short quarter-wavelength metal strips, as shown in Figure 9.5. This square is used to surround a patch antenna and research has shown that the outward propagating surface waves can be suppressed by such a configuration of RF shielding vias, thus alleviating the problem of diffraction at the edge of the substrate [12]. Additionally, it is also possible to build an integrated waveguide inside a substrate using such vias. Experiments have shown that metal filled via holes can improve isolation in LTCC RF multichip packages [13].
2. Signal vias. These are in effect through connections and terminate in each layer in cover pads. Via sizes are limited by the tape thickness, owing to the aspect ratio of filling the vias. Small-diameter, very deep vias are difficult to achieve as the filling of small deep holes is problematic. Keeping the aspect ratio of via to tape close to 1 is considered optimum for ease of manufacture.

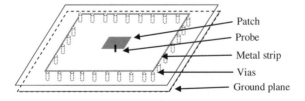

Figure 9.5 A patch antenna with an RF shielding via wall

3. Thermal vias. These features are for thermal conduction and are designed so that the connecting planes are staggered (offset) with adjacent planes at each layer, to maximise heat conduction through the volume of the material (see Figure 9.4).

9.2 Liquid Crystal Polymer

As an RF system's operating frequencies continue to rise, system reliability becomes increasingly reliant on hermetic or near-hermetic packaging materials. Higher frequencies lead to smaller circuits, and so low material expansion (which is related to water absorption) becomes more important for circuit reliability, as well as maintaining stable dielectric properties. Equally important is the ability to integrate these materials easily and cheaply with different system components. The best packaging materials in terms of hermeticity are: metals, ceramics and glass. However, nowadays these materials have often given way to cheaper polymer packages such as injection moulded plastics or glob top epoxies when cost is a concern [14]. Plastic packages are attractive from a cost point of view and for ease of fabrication, but they are not very good at excluding water and water vapour. A liquid crystal polymer (LCP) therefore offers a good combination of electrical, thermal, mechanical and chemical properties. Ideally, a hermetic polymer would be an inexpensive material, and generate low fabrication costs, while still functioning as a good microwave and millimetre wave package.

LCPs are organic materials that offer a unique all-in-one solution for high- frequency designs due to their ability to act as both a high-performance substrate and a packaging material for multilayer construction [15]. For gases, including oxygen, carbon dioxide, nitrogen, argon, hydrogen and helium, LCP also exhibits above-average barrier performance. Furthermore, the permeation of gases through LCP is not affected by humidity, even in an environment with elevated temperature (e.g. 150 °C) [16]. Very low water absorption (0.04 %) and high performance ($\varepsilon_r = 2.9$–3.0, tan $\delta = 0.002$–0.004) make LCPs very appealing for many applications, and are well placed as a prime technology for enabling system-on-package RF and millimetre wave designs [15].

LCP has been used as a microwave circuit substrate in thin-film form since the early 1990s when it was first recognised as a candidate for microwave applications [17–19]. However, early LCP films did not easily tear and were difficult to process. Lack of good film uniformity was not acceptable and poor LCP-to-metal adhesion, and failure to produce reliable plated through holes (PTHs) in LCP limited their capabilities for manufacturing circuits. Devising and optimising LCP surface treatments, via drilling and de-smearing techniques, were also necessary in order to bring the material into a state where circuits placed on it could be manufactured with confidence.

Liquid crystal polymers are identified as a class of thermoplastic polymer material with unique structural and physical properties. They contain rigid and flexible monomers that are linked to each other. When flowing in the liquid crystal state, rigid segments of the molecules align next to one another in the direction of shear flow (as in all liquid crystals). Once this orientation is formed, the direction of alignment and structure persist, even when the LCP is cooled below the melting temperature [20, 21]. This is different from most thermoplastic polymers (e.g. Kapton®), whose molecules are often randomly oriented in the solid state.

LCP almost satisfies the criteria for high-frequency design applications, so the material has attracted much attention for probably having the best packaging characteristics of all polymers. LCP has been called "near-hermetic" and has also been compared to glass in terms of water transmission. Previous literature has described the numerous benefits of LCP, including References [20] to [24]. These advantages are:

- Near-hermetic nature (water absorption 0.04 %) [25]
- Low cost
- Low-loss tangent (0.002 to 0.004 for 35 GHz)
- Low coefficient of thermal expansion (CTE), which may be engineered to match metals or semi-conductors
- Natural non-flammability (no need to add halogens, etc.)
- Recyclability
- Flexibility for conformal and/or flex circuit applications
- Excellent high-frequency electrical properties

Research work has also shown that solid-state devices such as pin diodes can be packaged in LCP [26], which offers a number of possibilities. In addition, several companies have recently developed injection- moulded LCP packaging caps [27, 28], which can be used to seal individual components with epoxy or laser sealing. However, these packages can be bulky, which may limit the packaging integration density. In addition, these rigid packaging "caps" (LCP becomes rigid when it has sufficient thickness) can take away one of the LCP substrates very unique characteristics – that of flexibility.

It is now necessary to look at the particular process of packaging devices in quantity with a standard thin-film LCP layer. For example, a 102 μm non-metallised LCP superstrate layer with depth-controlled laser micromachined cavities can be used. This technique has been demonstrated by creating packages for air-bridge RF MEMS switches. The switch membranes are only about 3 μm above the base substrate, which allows a cavity with sufficient clearance to be laser drilled in the LCP superstrate layer. A cavity depth of 51 μm (half of the superstrate thickness) was achieved for the MEMS package cavities.

This technique can be extended to include additional layers as necessary. To accommodate devices that require more vertical clearance, multiple LCP layers can have holes or cavities drilled in them and the layers stacked together (causing corresponding alignment issues). The packages can be sealed by thermocompression, ultrasonic or laser bonding.

At this point the RF characteristics of the discontinuity introduced by the LCP package cavities need to be discussed. A relatively new way of packing RF MEMS switches or MMICs has been reported in which multiple devices are located across an LCP substrate and the package fitted over both the transmission lines and the devices. Some of the advantages of this technique are: the flexibility of the substrate is maintained for applications such as conformal antennas, the package is lightweight, and the LCP packaging layer is a standard inexpensive microwave substrate which can be incorporated into a system-level package configuration. Two primary applications in this respect are large-scale antenna arrays with packaged ICs and/or switches within a multilayer antenna substrate, or perhaps vertically integrated LCP-based RF modules where switches and/or active devices may be bonded inside a multilayer LCP construction.

Since LCP has a low dielectric constant (near 3.16) [29], RF impedance mismatches are minimal when an LCP superstrate layer is added over a standard transmission line. Additionally, if cavities are machined in the superstrate layer, they do not create large impedance mismatches at the cavity interface. Thus, LCP's low dielectric constant offers the possibility for package cavities of arbitrary size to be integrated in a superstrate packaging layer accommodating chips, MEMS or other devices without concern for parasitic packaging effects. The LCP superstrate layer would then be bonded with a 25.4 μm thick, low melting temperature LCP bond layer to create an all-LCP package. The seal can then be created by the low melting temperature LCP (290 °C) layer, which has the same electrical characteristics as the high melting temperature (315 °C) core layers.

Figure 9.6 shows cross-sections of three different transmission lines. The first cross-section is a standard conductor-backed finite ground coplanar (CB-FGC) line, the second includes a 102 μm superstrate packaging layer and the third has a 51 μm laser machined cavity in the superstrate layer. The characteristic impedance difference of only 4 Ω between a transmission line with a superstrate layer compared to those with a cavity or without a packaging layer, means minimal reflections are created at the dielectric discontinuity.

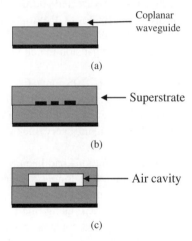

Coplanar
waveguide

(a)

Superstrate

(b)

Air cavity

(c)

Figure 9.6 Three different conductor-backed finite ground coplanar cross-sections present in the measured packaging structures: (a) without superstrate, (b) with superstrate and (c) with a LCP package

LCP's low water absorption makes it stable across a wide range of environments by preventing changes in the relative dielectric constant and loss tangent. LCP material processing is still in its infancy, and its materials cost is of the same magnitude as those in Table 9.1.

Table 9.1 Material comparison

	ε_r	tan δ	Operating frequency (GHz)	Ref.
LCP	2.9–3.2	0.0020–0.0045	<105	[29]
LTCC	5.7–9.1	0.0012–0.0063	<65	[1]
FR4	4.4	0.025	<10	[29]

However, due to the capability of LCP to be handled by reel-to-reel processing, it is expected that production costs will continue to fall. At the same time, the material's flexibility and relatively low processing temperatures make it suitable for applications such as conformal antenna arrays, and for the integration of microelectromechanical system (MEMS) devices, such as low-loss RF switches.

As will be realised from the above, multilayer LCP circuits are feasible due to the material features of different melting temperatures. High melting temperature LCP (315 °C) can be used for core layers, while low melting temperature LCP (290 °C) is used as a bond ply. Thus, vertically integrated designs may be realised that are similar to those in LTCC. An additional benefit in multilayer LCP fabrication is the functionality provided by the low dielectric constant. This is useful for vertically integrated designs where the antenna is printed on the top layer of an all-LCP module.

The fabrication difficulties of LCP have been solved gradually over the years [30–37]. A biaxial die extrusion process was developed [19, 25] that solved the tearing problems by giving the material uniform strength and it also created additional processing benefits. It was discovered that by controlling the angle and rate of LCP extrusion through the biaxial die, the *xy* coefficient of thermal expansion (CTE) could be controlled to between approximately 0 and 40 ppm/°C. Thus, by this process a thermal expansion match in the *xy* plane can be achieved with many commonly used materials.

LCP's *z* axis CTE is considerably higher, ~ 105 ppm/°C , but due to the thin layers of LCP used, the absolute *z* dimension difference between LCP and a 51 μm high copper plated through holes, is less than one of a half-micrometre within a ± 100°C temperature range [38]. This makes *z* axis expansion a minimal concern until very thick multilayer modules come into consideration.

Since 2002, many of the LCP process limitations had been overcome [39], and it became commercially available in thin films with single and double copper cladding. Interest has grown quickly in utilising LCP for higher-frequency applications [40, 41]. Many publications [42, 43] have reported microwave characterization of LCP using microstrip ring resonators of up to 34.5 GHz. Additionally, a 50 Ω conductor-backed coplanar waveguide (CB-CPW) transmission line on LCP [44], has shown LCP to have low loss from 2 to 110 GHz, and a coplanar waveguide (CPW) on LCP [45] has been measured to 50 GHz.

However, achieving broadband dielectric material characterisation at higher frequencies is not a trivial task. The ring resonator (see Figure 9.7) provides dielectric information at discrete frequency points at periodic resonant peaks, but substrate thickness, ring diameter and the dielectric constant of the material under test all affect the accuracy of the measurement. In addition, at high frequencies where the skin depth approaches the characteristic dimension of the surface roughness of the resonator's metal lines, it becomes difficult to separate the effects of the conductor from dielectric losses.

Figure 9.7   Ring resonator

Multiple dielectric characterisation methods can be performed to accurately identify dielectric properties of LCP for frequencies from 30 to 110 GHz [29]. Microstrip ring resonators (Figure 9.7) of varying diameters and substrate thicknesses, cavity resonators and transmission line methods have all been used and experimentally cross-referenced to determine accurately the wideband characteristics of LCP. In addition, coplanar waveguides (CPWs) and microstrip lines, each on varying substrate thicknesses, have been investigated from 2 to 110 GHz and the losses have been quantified. These transmission line losses across the millimetre wave range can provide design guides for loss versus frequency of circuits built on LCP substrates. The results of these measurements have yielded a thorough knowledge of LCP dielectric properties and the performance of LCP-based circuits in millimetre wave RF systems.

LCP was originally used as a high-performance thermoplastic material for high-density printed circuit board (PCB) fabrication [16, 46] and semiconductor packaging [47], but subsequent research work has shown that LCP is virtually unaffected by most acids, bases and solvents over a considerable period of time and over a broad temperature range [16]. In terms of mechanical properties, the thermal expansion coefficient of the LCP material can be controlled during the fabrication process to be both small and predictable [47]. For LCP films, with uniaxial molecular orientation, its mechanical properties will be anisotropic and dependent on the orientation of the polymer chains. As an illustration, uniaxial LCP film can withstand less load in the transverse direction (i.e. the direction orthogonal to the orientation of its molecular chains), than in the longitudinal direction (i.e. the direction along the orientation of its molecules) [23].

To overcome this problem, biaxially oriented film with uniform transverse and longitudinal direction properties can be fabricated. The orientation of LCP molecules varies through the thickness of the film, while at the two faces of the film molecules are oriented orthogonally. When the orientation angles are $+45°$ and $-45°$ to the stress direction on each surface, the mechanical properties, such as the coefficient of thermal expansion, tensile strength and modulus, are almost isotropic [47].

Commercial LCP material is supplied in thin film with predefined thicknesses ranging from 25 μm to 3 mm. One or both sides of the LCP film can have 18 μm thick copper cladding. This copper layer is laminated in a vacuum press at a temperature around the melting point of LCP [47]. The copper layer can be used for multilayer antenna construction with bonding films.

There are many types of LCP products and their properties vary slightly between types. It is useful to compare LCP with Kapton® [48], a polymer film that has been used in MEMS in recent years. Compared with Kapton® polyimide film (Table 9.2), LCP has a low cost (about 50–80 % lower than Kapton®), is relatively unaffected by moisture and humidity, is not attacked by certain caustic solutions [16] and is amenable to melt processing. As a result, bonding between LCP and other substrates (e.g. glass) is simplified. For example, whereas Kapton® is often bonded with an intermediate adhesion layer, LCP films can bond to other surfaces directly by thermal lamination.

Table 9.2 Comparison of LCP and Kapton®

	Dielectric constant	Loss tangent	Tensile strength (kpsi)
LCP	2.8	∼0.004	30
Kapton®	3.5	∼0.002	34

It is possible to change the permittivity of nematic liquid crystals when a DC voltage is applied. One example of an application exploiting this property, is a tuneable antenna element with LCP at millimetre wave frequencies [20]. The concept is based on modifying the resonant frequency of a microstrip patch by changing the permittivity of the LCP. In order to achieve this, a cavity has to be formed under the patch filled with LCP, as shown in Figure 9.8.

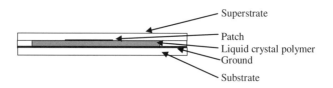

Figure 9.8   Configuration of an LCP-loaded cell employing a microstrip patch element

9.3 CMOS

Historically, monolithic microwave integrated circuits (MMICs) have been designed using III–V semiconductor technologies such as GaAs and InP. These have superior performance compared to CMOS due to their higher electron mobility, higher breakdown voltage and the availability of high-quality factor passive components for this technology. However, a CMOS implementation offers higher levels of integration and reduced cost. Several recent developments have combined to enable CMOS circuit blocks to operate at very much higher frequencies compared to earlier approaches. Firstly, millimetre wave CMOS circuits directly benefit from the higher speed of the scaled technology. Additionally, improved circuit topologies and new design approaches have been introduced to fully exploit the intrinsically faster devices. At present, only CMOS oscillators [47–51]have been demonstrated beyond 30 GHz, while CMOS amplifiers [52–54] and mixers [54, 55] have only achieved operation up to 26 and 21.8 GHz. A key reason for this large discrepancy of operating frequencies, is the lack of accurate CMOS active and passive device models at millimetre wave frequencies. Thus circuits employing these components will need to wait for these more accurate models to become available.

The substrate resistivity of most modern standard silicon processes is $10\,\Omega$-cm, which is many orders of magnitude lower than that of GaAs (10^7–$10^9\,\Omega$-cm) [56]. Signals that couple to a low-resistivity silicon substrate incur significant losses, especially at millimetre wave frequencies. However, the gate material used for CMOS devices is polysilicon, which has a much higher sheet resistance ($\sim 10\,\Omega$/square) than the metal used for the gates of GaAs field effect transistors (FETs). A higher gate resistance can reduce the transistor power gain and increase noise. Fortunately, simple layout techniques can be used to minimise the detrimental effects of the polysilicon gate [57].

The low substrate resistance of CMOS has a direct consequence on the design of transmission lines while the gate resistance can affect the design of switches; both are important structures for millimetre wave design. At these frequencies, the reactive elements needed for matching networks and resonators become increasingly small, requiring inductance values of the order of 50–250 pH. Given the quasi-transverse electromagnetic (quasi-TEM) mode of propagation, transmission lines are inherently scalable in length and are capable of realising the precise values of small reactances. As a further attraction, interconnect wiring can be modelled directly when implemented using transmission lines. Another benefit of using transmission lines is that the well-defined ground return path, significantly reduces magnetic and electric field coupling to adjacent structures.

Any quasi-TEM transmission line can be characterised using its equivalent frequency-dependent distributed circuit model (Figure 9.9). The transmission line can also be characterised by the following four real parameters:

$$Z \equiv \sqrt{L/C}$$

$$\lambda \equiv \frac{2\pi}{\omega_0 \sqrt{LC}} \qquad (9.1)$$

$$Q_L \equiv \omega_0 L / R$$

$$Q_C \equiv \omega_0 C / G$$

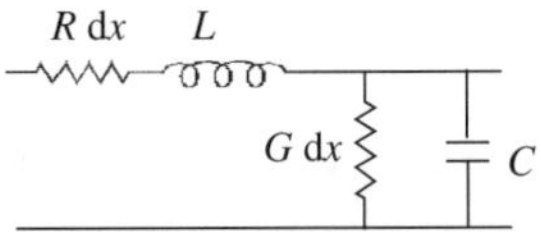

Figure 9.9 Distributed model for a lossy transmission line

Transmission lines implemented on GaAs have no shunt loss, but transmission lines implemented on low-resistivity silicon often have low capacitive quality (Q_C) factors due to substrate coupling. For transmission lines that store essentially magnetic energy, the inductive quality factor (Q_L) is the most critical parameter when determining the loss of the line, as opposed to the resonator quality factor, or the attenuation constant.

Microstrip lines on silicon are typically implemented using the top-layer metal as the signal line and the bottom-layer metal for the ground plane. Figure 9.10 (a) illustrates the effectiveness of the metal shield, with essentially no electric field penetration into the substrate. The shunt loss, G, is therefore merely caused by the loss tangent of the oxide, yielding a capacitive quality factor, Q_c, of around 30 at millimetre wave frequencies [57]. The main disadvantage to microstrip lines on standard CMOS is the close proximity of the ground plane to the signal line (~ 400 nm), yielding a very small distributed inductance, L. This significantly degrades the inductive quality factor, Q_L [57].

An alternative design for on-chip transmission lines is the use of coplanar waveguides (CPWs) [58, 59], which are implemented with one signal line surrounded by two adjacent grounds (in the same plane, see Figure 9.10 (b)). The signal line width is used to minimise conductor loss, while the signal-to-ground spacing, controls the impedance and hence the trade-off between Q_L and Q_C. For instance, a CPW of 10 μm width and 7 μm space has a 59Ω characteristic impedance, and a Q_L measured to be about double that of the microstrip [57]. Therefore, CPW transmission lines have considerably higher Q_L compared to microstrip lines and so are commonly used in millimetre wave design. It is also easy to take measurements using wafer probes (Figure 9.10 (c)).

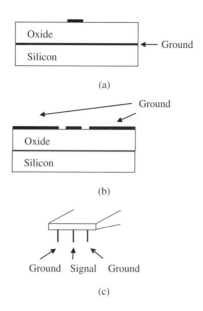

Figure 9.10   (a) Microstrip, (b) coplanar waveguide transmission lines and (c) wafer probe

Microstrip lines have, to a first-order approximation, constant Q_L and Q_C regardless of geometry. Another important issue when designing with CPWs is the unwanted odd (asymmetric) CPW mode, which arises because CPW lines inherently have three conductors. To suppress this parasitic propagation mode, the two grounds should be forced to the same potential [58]. In MMICs, this requires the availability of air bridge technology, which is costly. But underpasses using a lower metal level in a modern CMOS process can be used to suppress this mode.

In a typical CMOS circuit, the substrate losses seriously limit the Q-factor of a conventional microstrip line, or a CPW above the CMOS SiO_2 layer, to around between six and ten, or even higher. In the CMOS process, much thinner metal layers than those incorporated in III–V-based ICs are adopted in typical standard foundry processes. This makes improvement of the Q-factor of CMOS transmission lines a difficult task. However, if the back plating layer of a typical CMOS microstrip can be raised to the plane of one of the multimetal layers of the CMOS technology, the resultant new thin-film microstrip will exhibit a smaller cross-section. Thus microstrip circuits of much smaller widths would have the same characteristic impedance as that of a much wider microstrip on a CMOS substrate. The penalty for doing this is the substantial increase in the attenuation constant, which can inhibit electrical performance. Despite this drawback, a thin-film microstrip is still appealing [60], since a range of small passive components can be integrated.

Figure 9.11 [61] illustrates the unit cell of one example incorporating the complementary conducting strip transmission line. On the bottom layer is the connected unit cell, and on the top surface the unit cell is patterned to make a 50Ω transmission line. When viewed in cross-section, the unit cell comprises a raised CPW and a thin-film microstrip transmission line connected in series. The ground planes of the raised CPW and that of the thin-film microstripline are connected via the bottom metallic plane. The composite transmission line is a succession of

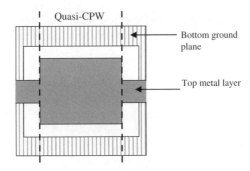

Figure 9.11   Top view of a unit cell of the complementary conducting strip [61]

such unit cells in series, which may have linear signal paths bent to achieve the desired signal routing. Therefore these cells constitute a complete guiding structure in a periodic arrangement. When viewed from the top surface, the upper metal layer and bottom metal surface complement each other. The complementary conducting strip (CCS) (Figure 9.11) transmission line is a periodic array, along which the guided wave experiences a perturbation in the longitudinal direction. It will be observed that this guiding structure is two dimensional, which increases the degrees of freedom for designing transmission lines of arbitrary characteristic impedance and current handling capability. Published work indicates that a very compact layout of the transmission line, equivalent to that of the LC tank circuit, can be implemented as the adjacent cells have negligible coupling (typically well below coupling of -20 dB [61]). This type of transmission line has been implemented in a printed rat-race hybrid at 39 GHz on CMOS [62] and can also be used in a millimetre wave feeding network.

9.4 Meta Materials

The concept of metamaterials, was first postulated in the late 1960s, but only recently has the subject attracted significant interest since practical implementation solutions have emerged. Metamaterials are man-made composite structures with artificial elements (much smaller than the wavelength of electromagnetic propagation) situated within a carrier medium. These materials can be designed with arbitrary permeability and permittivity [63]. Left-handed materials are characterised by a negative permittivity and a negative permeability- at least across a portion of the electromagnetic frequency spectrum. As a consequence, the refractive index of a metamaterial can also be negative across that portion of the spectrum. In practical terms, materials possessing such a negative index of refraction are capable of refracting propagating electromagnetic waves incident upon the metamaterial in a direction opposite to that of the case where the wave was incident upon a material having a positive index of refraction (the inverse of Snell's law of refraction in optics). If the wavelength of the electromagnetic energy is relatively large compared to the individual structural elements of the metamaterial, then the electromagnetic energy will respond as if the metamaterial is actually a homogeneous material. Published work shows that antenna gain can be enhanced by using metamaterials as antenna substrates [64].

As these materials can exhibit phase and group velocities of opposite signs and a negative refractive index in certain frequency ranges, both of these characteristics offer a new design concept for RF and microwave applications. One of the approaches starts from the equivalent transmission line model and artificially loads a host line with a dual periodic structure consisting of series capacitors and shunt inductors [65]. The length of the period and the values of the capacitors and inductors determine the frequency band in which the material has this double negative behaviour. One of the challenges for implementation of these concepts for very high frequencies, where the dimensions of the components become smaller and the process design rules become very restrictive, is the choice of the inductor and capacitor geometry to obtain the required left-handed passband and minimum insertion loss at the desired operating frequency [65].

One of the implementations of such structures is in arrays of wires and split-ring resonators [66, 67]. These three-dimensional structures are complicated and are difficult to apply to RF and microwave circuits. A more practical implementation uses transmission lines periodically loaded with lumped element networks [68, 69]. The starting point is the transmission line model presented in Figure 9.12(a).

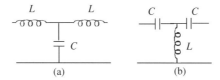

Figure 9.12   (a) *L-C-L* and (b) *C-L-C* transmission line models [65]

The equivalence between the distributed L and C for the transmission line and the permittivity and permeability of the medium is expressed as $\varepsilon = C$, $\mu = L$. By periodically loading this transmission line with its dual in Figure 9.12 (b), the values of ε and μ change as follows [68]:

$$\varepsilon_{\text{eff}} = \varepsilon - \frac{1}{\omega^2 L d} \quad \mu_{\text{eff}} = \mu - \frac{1}{\omega^2 C d} \tag{9.2}$$

where ε and μ are the distributed inductance and capacitance of the host transmission line repectively. It is clear from Equation (9.2) that for certain values of L, C and d, the effective permittivity and permeability of the medium becomes negative for some frequency ranges. In these ranges, the refractive index is negative and the phase and group velocities have opposite signs.

The example structure given in Figure 9.13 can be implemented in many dielectric substrates, and the host transmission line can be a 75 Ω, coplanar waveguide (CPW). The advantage of the CPW is the ease of building shunt lumped elements, due to the availability of the ground plane on the same layer as the signal, thus eliminating the need for vias. The series capacitors and shunt inductors can be implemented as shown in Figure 9.13.

Much research has been accomplished regarding the manufacture, the properties and the applications of metamaterials. Figure 9.14 shows a top view and a three-dimensional view of illustrative metamaterial structures that can be used in antenna arrays. The metamaterials in Figure 9.14 are of the type investigated by Caloz *et al.* [70].

In Figure 9.14, three unit-cell circuit structures are repeated periodically along the microstripline. A unit-cell circuit, in the structure, consists of one or more electrical components that are

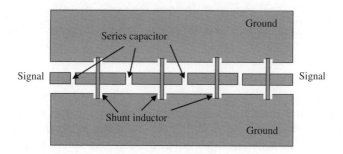

Figure 9.13   CPW implementation of a metamaterial structure

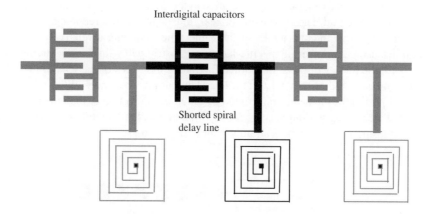

Figure 9.14   Periodic structure of three unit cells

repeated- in this case disposed along the microstrip transmission line. In the structure in Figure 9.14 above, series interdigital capacitors are placed periodically along the line and T-junctions between each of the capacitors connect the microstripline to shorted spiral stub delay lines that are, in turn, connected to ground by vias. The microstrip structure of one capacitor, one spiral inductor and the associated ground via, form the unit-cell circuit structure of Figure 9.14.

Structures similar to Figure 9.14 can be used in leaky-wave antennas (as opposed to phased-array antennas), which have been designed to operate at frequencies of up to approximately 6.0 GHz [70]. With certain modifications, these metamaterials can be used at relatively high frequencies, such as those frequencies useful in millimetre wave communications applications i.e. above 60 GHz [71]. For instance, the unit-cell circuit structure of Figure 9.14 can be reduced to a size much smaller than the effective wavelength of the signal. To achieve a high-performance transmission line impedance at a particular frequency, the physical size and positioning of unit cells in the metamaterial microstripline needs to be carefully considered.

High-gain printed arrays have previously relied on a signal-feed/delay line architecture that resulted in a biconvex, or Fresnel, lens for focusing the microwaves [71]. The use of such lens architectures has resulted in microwave radiation patterns having relatively poor sidelobe performance due to attenuation as the wave passed through the lens. Specifically, the signal passing through the central portion of the lens tended to be attenuated to a greater degree than the signal passing through the edges of the lens. This resulted in an aperture distribution function that was "darker" in the centre of the aperture and "brighter" near the edges. The diffraction pattern of this function results in significant sidelobes (the diffraction or far-field radiation pattern is the two-dimensional Fourier transform of the aperture distribution function). While placing signal delay lines in the lens portion of the system could reduce the sidelobes and, as a result, increase the performance of a phased-array system, this was deemed to be limited in its usefulness because, by including such delay lines, the operating bandwidth of the phased-array system was reduced. However, instead of a biconvex lens, a metamaterial can be used to create a biconcave lens (by means of controlling the effective refractive index of the material) for focusing the wave transmitted by the antenna. As a result, a wave passing through the centre of the lens is attenuated to a lesser degree relative to the edges of the lens (the aperture is now brighter at the centre and darker near the edges), thus significantly reducing the amplitude of the sidelobes of the antenna while, at the same time, retaining a relatively wide useful bandwidth. By using a number of specific-length microstrip delay lines in the architecture as in Figure 9.15, the phases of the signals travelling along the edges of the lens can be delayed relative to those travelling in the centre of the lens. Thus, as previously described, the amplitude of the central portion of the beam transmitted by antenna can be higher than the amplitude at the edges and, accordingly, sidelobes are reduced (of course, the effective aperture efficiency needs to be considered if the gain is an important design factor).

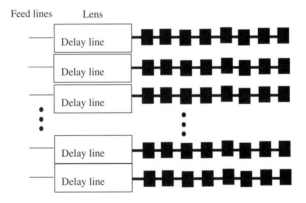

Figure 9.15   Beam-steering array

Metamaterial delay lines can be realised as lines lithographically printed on a suitable substrate. One or more electronic components, such as amplifiers, may be deposited along each of the delay lines. The delay lines thus form an electromagnetic lens that is used to delay and/or amplify the individual signals travelling along each delay line. Such (variable) delay lines can also be used to steer and focus the beams produced by the antenna. However, as will be realised, delay lines also reduce the useful bandwidth of the phased array antenna system.

9.5 High-Temperature Superconducting Antenna

High-temperature superconducting antennas can exhibit an increase in overall gain compared to their copper equivalent [72]. Many designs have been published and structures realised, as microstrip arrays directly corresponds to the conventional copper designs. At millimetre wave frequencies, transmission line loss is important in determining the feasibility of an antenna array. When a resonant array employs a corporate feed network, an additional effective path length is introduced so the actual loss will unavoidably increase. Travelling wave arrays have shorter effective path lengths than resonant arrays and can appear to offer less loss. In a real implementation for waveguide arrays, however, the actual loss will be greater, due to surface roughness, metal imperfections, etc. Use of superconducting waveguides can allow the efficiency of the feed component of the antenna to approach 0 dB, and thus may allow a significant extension of array techniques. For a printed array, it has been reported that a microstrip of conventional dimensions experiences a loss reduction of the order of 20 dB at 100 GHz, while a thin-film microstrip realisation could show another 10 dB of loss reduction [73]. Thus with the reduction in resistive loss associated with superconductors, high-gain microstrip arrays become a practical proposition at millimetre wave frequencies.

The use of high-temperature superconductor (HTS) thin-film structures in microwave integrated circuits employed in aerospace applications and terrestrial mobile communication base stations, offers the possibility of significantly reducing the weight and volume of the microwave equipment, even though the system must include cryogenic equipment to provide an operational temperature in the range of 60 to 77 K. Thus it can be said that the HTS applications in microwave engineering have become part of common industrial practice [74]. This emphasizes the importance of investigation into the physical properties of HTS thin films and of the development of microwave characteristics of such films. Thin HTS films are prepared by epitaxial growth on single-crystal dielectric substrates. In many cases, the film quality is governed to a considerable extent by the processes occurring at the interface between the film and the substrate. As a consequence, the state of the interface in heteroepitaxial systems consisting of HTS films and dielectric substrates becomes extremely important from the standpoint of practical applications of HTS in modern microwave electronics [74].

Superconductivity was initially discovered in 1911 and the highest superconducting transition temperature observed was in the Nb_3Ge compound (23.2 K), and this was the highest transition temperature known of until 1986, when high-temperature superconductivity in the La–Ba–Cu–O ceramic was announced. Before high-temperature superconductivity had been established to exist, oxide-type superconductors were considered a curiosity and not worthy of serious consideration.

At the beginning of 1987, the situation changed abruptly, and the HTS became a subject of worldwide interest. Figure 9.16 depicts the history of high-temperature superconductors in terms of the compounds and their superconducting transition temperature. As the initial interest gradually abated, the area of HTS applications has become the major focus of research. Although the detailed physics of high-temperature superconductors differs from that of the earlier (low-temperature) superconductors, they nevertheless possess many of the properties that are of interest to many applications in engineering. The $YBa_2Cu_3O_{7-x}$ and $Tl_2Ba_2Ca_2Cu_3O_{10}$ compounds deposited as thin films were found suitable for practical applications as a basis for microwave devices [75, 76], with transition temperatures relatively easily attained, both in the laboratory and in the field, for commercial applications.

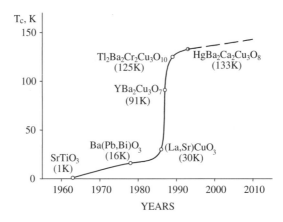

Figure 9.16 The history of high-temperature superconductors [77]

While HTS materials for research are currently grown in the form of perfect single crystals and these are used in investigations of the fundamental properties of substances, no commercial devices have yet been designed on the basis of HTS single crystals. The use of HTS in electronics is mainly based on polycrystalline thin films on a (low-loss) dielectric substrate. In the case of microwave applications, the best substrate for $YBa_2Cu_3O_{7-x}$ films was found to be sapphire buffered by a thin CeO_2 layer [78, 79]. One example of a superconducting dipole antenna is shown in Figure 9.17. The dipole is fed by means of a feed line through a via. However, the problem of growing perfect HTS epitaxial films has not been solved completely. Process optimisation for film integrity is a skill akin to many processes requiring a high degree of expertise in the semiconductor industry. Processing, materials, conditions and film properties all have to be controlled reliably for consistent films to be produced. In particular, the properties of the superconductor can vary according to the direction of the crystallographic axes (termed a, b and c axes). These superconductors have a structure that is termed a Perovskite lattice (from the naturally occurring mineral). Figure 9.18 illustrates the formation of $YBa_2Cu_3O_{7-x}$ grains on a crystalline film, where (i) is an epitaxial c-axis HTS grain, (ii) is an a-axis HTS grain and (iii) is an axially misaligned HTS grain.

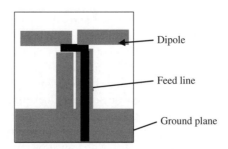

Figure 9.17 $YBa_2Cu_3O_{7-x}$ dipole antenna

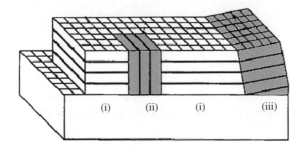

Figure 9.18 Possible formation of YBa$_2$Cu$_3$O$_{7-x}$ grains on a crystalline substrate

The surface resistance of HTS at microwave frequencies is responsible for the loss in planar antenna feeding lines and for the decay of oscillation in resonators. Therefore discussed next will be the role of this property in surface impedance.

The surface impedance of an HTS material for a plane electromagnetic wave incident normally to its surface is defined as the ratio of $|E|$ to $|H|$ on the surface of the sample. It is described by the equation:

$$Z_{sur} = R_{sur} + iX_{sur} = \left(\frac{i\omega\mu_0}{\sigma_1 - i\sigma_2}\right)^{1/2} \tag{9.3}$$

where R_{sur} and X_{sur} are the surface resistance and the surface reactance, respectively, $\omega = 2\pi f$, f is the frequency in Hz, μ_0 is the magnetic permeability of free space, and σ_1 and σ_2 are the real and imaginary parts of the conductivity.

The two-fluid model proposed by Gorter and Casimir [80] is commonly used for a realistic description of the HTS surface impedance [75, 76]. In accordance with this model, the components of the conductivity can be written as:

$$\sigma_1 = \frac{e^2 n_n \tau}{m} \frac{1}{1 + (\omega\tau)^2} \tag{9.4}$$

$$\sigma_2 = \frac{e^2 n_s}{\omega m} \left[1 + \frac{n_n}{n_s} \frac{\omega\tau}{1 + (\omega\tau)^2}\right] \tag{9.5}$$

where e and m are the charge and the effective mass of the electron respectively, τ is the relaxation time, and n_n and n_s are the densities of the normal and superconducting charge carriers.

Within the microwave and millimetre wave frequency range [$(\omega\tau)^2 << 1$], Equations (9.4) and (9.5) can be simplified to:

$$\sigma_1 = \frac{e^2 n_n \tau}{m} \tag{9.6}$$

$$\sigma_2 = \frac{1}{\omega\mu_0\lambda_L^2} \tag{9.7}$$

where:

$$\lambda_L = \left(\frac{e^2 n_s \mu_0}{m} \right)^{-1/2}$$

where λ_L is the London penetration depth [75].

Substituting Equations (9.6) and (9.7) into Equation (9.3) and taking into account that for $T < T_c$ the inequality $\omega \mu_0 \sigma_n \lambda_L^2 << 1$ is valid, it can be stated that [75, 76]:

$$R_{sur} = \frac{1}{2} (\omega \mu_0)^2 \sigma_n \lambda_L^3$$

$$X_{sur} = \omega \mu_0 \lambda_L$$

The parameters σ_n and λ_L are temperature dependent. It should be noted that surface resistance varies as $f^{1/2}$ for normal metals, and f^2 for superconductors. The frequency at which the superconductor's surface resistance becomes equal to the surface resistance of copper is called the crossover frequency. Experiments show that the crossover frequency for high-temperature superconductors such as YBCO and copper is above 100 GHz [77]. Therefore, below this frequency, HTS materials have an advantage over copper and normal conductors at the same temperature. A comparison of surface resistance between TBCCO superconductor thin film and copper is plotted in Figure 9.19. This shows that a superconducting film has less than one-tenth of the surface resistance of copper.

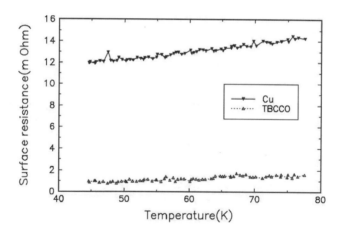

Figure 9.19   Surface resistance comparison between TBCCO and Cu at 40 GHz [81]

For a superconducting antenna such as a helix, the antenna efficiency is then given by [72]:

$$\eta = \frac{R_r}{R_r + R_l + R_m} \tag{9.8}$$

where R_r is radiation resistance, R_l is loss resistance, and R_m is an equivalent resistance associated with losses in the matching network. R_l is proportional to surface resistance. Therefore, antenna efficiency can be improved as the surface resistance is reduced, and superconducting antennas could have better efficiency than copper antennas at the same temperature.

In addition to millimetre wave antenna elements, high-temperature superconductors can also be applied to superior-performance millimetre wave feeding networks, millimetre wave resonators and interconnects. The use of high dielectric constant substrates also means that components at these high frequencies can be considerably smaller than in conventional circuits. Metamaterial components can also be realised, and bulk components have begun to appear as research subjects. The advantages of high-temperature superconductors in millimetre wave applications of up to 100 GHz are many, and although early commercial products at microwave frequencies exist (e.g. in cellular base stations), the investigation into the use of these materials at these higher frequencies remains a laboratory activity.

9.6 Nano Antennas

Nanotubes have different characteristics from conventional metallic transmission lines. For instance, the wave velocity on a carbon nanotube transmission line in free space is of the order of the Fermi velocity v_F, rather than the speed of light c. For a carbon nanotube, v_F is about 9.71×10^5 m/s. It is found that the propagation velocity on a carbon nanotube dipole is $v_p \approx 6.2 v_F \approx 0.02c$. Thus, wavelengths are much shorter in a carbon nanotube, compared to those in a conventional microscopic metallic tube.

Basic properties of dipole transmitting antennas formed by carbon nanotubes have been studied [82]. Some properties of carbon nanotube antennas are found to be quite different from the case of an infinitely thin copper antenna of the same size and shape. It is found that carbon nanotube antennas have plasmon resonances above a critical frequency, have high input impedances (which can be used for connecting to nanoelectronic circuits), and have very low efficiencies.

A model for parallel wire carbon nanotube transmission lines has been created [83]. Surface waves on carbon nanotubes and carbon nanotube antennas are discussed there. Using carbon nanotubes, slow-wave structures with a wave velocity reduction of two orders of magnitude can be realised. This opens up the potential for ultra-compact resonant antennas and extremely small distributed circuits, e.g. surface wave filters for frequencies of GHz, and up into the THz range [84].

References

[1] A. Matsuzawa, RF-SoC – Expectations and Required Conditions', *IEEE Transactions on Microwave Theory and Techniques*, **50**, January 2002, 245–253.

[2] K. Lim, S. Pinel, M. F. Davis, A. Sutono, C.-H. Lee, D. Heo, A. Obatoynbo, J. Laskar, M. Tentzeris and R. Tummala, 'RF-SOP for Wireless Communications', *IEEE Microwave Magazine*, **3**, March 2002, 88–99.

[3] T. Kutilainen, 'Ceramic Interconnect Initiative. NextGen, LTCC', ELMO, November 2005, http://www. imaps.org/cii/

[4] Young Chul Lee, Won-il Chang and Chul Soon Park, 'Monolithic LTCC SiP Transmitter for 60GHz Wireless Communication Terminals', *IEEE Transactions on Microwave Theory and Techniques*, **MTT-S**, June 2005, 1015–1018.

[5] K. Lim, S. Pinel, M. F. Davis, A. Sutono, C.-H. Lee, D. Heo, A. Obatoynbo, J. Laskar, E. M .Tentzeris and R. Tummala, 'RFSystem-on-Package (SOP) for Wireless Communications', *IEEE Microwave Magazine*, **3**(1), March 2002, 88–99.

[6] J. Lee, K. Lim, S. Pinel, G. DeJean, R. L. Li, C.-H. Lee, M. F. Davis, M. Tentzeris and J. Laskar, 'Advanced System-on-Package (SOP) Multilayer Architectures for RF/Wireless Systems up to Millimetre-Wave Frequency Bands', Proceedings of the Asian Pacific Microwave Conference, Seoul, Korea, November 2003, pp. FA5_01.

[7] R. Lucero, W. Qutteneh, A. Pavio, D. Meyers and J. Estes, 'Design of an LTCC Switch Diplexer Front-End Module for GSM/DCS/PCS Application', IEEE Radio Frequency Integrated Circuit Symosium, Phoenix, Arizona, May 2001, pp. 213–216.

[8] Keko, 'LTCC Production', Keko Equipment Newsletter, November 2003, No. 2, pp. 6–7.

[9] Young Chul Lee and Chul Soon Park, 'A Novel High-Q LTCC Stripline Resonator for Millimetre-Wave Applications', IEEE Microwave and Wireless Components Letters, 13(12), 2003, 499–504.

[10] S. Maci, L. Borelli and L. Rossi, 'Diffraction at the Edge of a Truncated Grounded Dielectric Slab', IEEE Transactions on Antennas and Propagation, 44, June 1996, 863–872.

[11] D. Moongilan, 'Grounding Optimization Techniques for Controlling Radiation and Crosstalk in Mixed Signal PCBs', Proceedings of the IEEE Electromagnetic Compatibility Symposium, 1, August 1998, 495–500.

[12] G. DeJean, R. L. Li, M. M. Tentzeris, J. Papapolymerou and J. Laskar, 'Radiation-Pattern Improvement of Patch Antennas Using a Compact Soft/Hard Surface (SHS) Structure on LTCC Multilayer Technology', IEEE Transactions on Antennas and Propagation, AP-S, 2004, 317–320.

[13] George E. Ponchak, Donghoon Chun, Jong-Gwan Yook and Linda P. B. Katehi, 'The Use of Metal Filled Via Holes for Improving Isolation in LTCC RF and Wireless Multichip Packages', IEEE Transactions on Advanced Packaging, 23(1), February 2000, 88–99.

[14] Dane Thompson, Nickolas Kingsley, Guoan Wang, John Papapolymerou and Manos M. Tentzeris, 'RF Characteristics of Thin Film Liquid Crystal Polymer (LCP) Packages for RF MEMS and MMIC Integration', IEEE Transactions on Microwave Theory and Techniques, MTT-S, June 2005.

[15] S. Pinel, M. Davis, V. Sundaram, K. Lim, J. Laskar, G. White and R. Tummala, 'High Q Passives on Liquid Crystal Polymer Substrates and μBGA Technology for 3D Integrated RF Front-End Module', IEICE Transactions on Electronics, E86-C(8), August 2003, 1584–1592.

[16] E. C. Culbertson, 'A New Laminate Material for High Performance PCBs: Liquid Crystal Polymer Copper Clad Films', Proceedings of the 45th Electronic Components and Technology Conference, 1995, pp 520–523.

[17] K. Jayaraj, T. E. Noll and D. R. Singh, 'RF Characterization of a Low Cost Multichip Packaging Technology for Monolithic Microwave and Millimetre Wave Integrated Circuits', URSI International Signals, Systems, and Electronics Symposium, October 1995, pp. 443–446.

[18] E. C. Culbertson, 'A New Laminate Material for High Performance PCBs: Liquid Crystal Polymer Copper Clad Films', IEEE Electronic Components and Technology Conference, May 1995, pp. 520–523.

[19] K. Jayaraj, T. E. Noll and D. R. Singh, 'A Low Cost Multichip Packaging Technology for Monolithic Microwave Integrated Circuits', IEEE Transactions on Antennas and Propagation, 43, September 1995, 992–997.

[20] R. Marin, A. Mössinger, J. Freese, A. Manabe and R. Jakoby, 'Realization of 35 GHz Steerable Reflectarrays Using Highly Anisotropic Liquid Crystal', IEEE Transactions on Antennas and Propagation, AP-S, 2006, 4307–4310.

[21] L. Chorosinski, 'Low Cost, Lightweight, Inflatable Antenna Array Development Using Flip Chip on Flexible Membranes for Space-Based Radar Applications', Northrop Grumman Company Report, 2000

[22] H. Inoue, S. Fukutake and H. Ohata, 'Liquid Crystal Polymer Film Heat Resistance and High Dimensional Stability, Proceedings of the Pan Pacific Microelectronics Symposium, February 2001, pp. 273–278.

[23] X. Wang, J. Engel and C. Liu, 'Liquid Crystal Polymer (LCP) for MEMS: Processes and Applications', Journal of Micromechanical Microengineering, 13, September 2003, 628–633.

[24] L. M. Higgins III, 'Hermetic and Optoelectronic Packaging Concepts Using Multiplayer and Active Polymer Systems', Advancing Microelectronics, 30(4), July/August 2003, 6–13.

[25] B. Farrell and M. St Lawrence, 'The Processing of Liquid Crystalline Polymer Printed Circuits', IEEE Electronic Components and Technology Conference, May 2002, pp. 667–671.

[26] G. Zou, H. Gronqvist, J. P. Starski and J. Liu, 'Characterization of Liquid Crystal Polymer for High Frequency System-in-a-Package Applications', IEEE Transactions on Advanced Packaging, 25, , November 2002, 503–508.

[27] K. Gilleo, J. Belmonte and G. Pham-Van-Diep, 'Low Ball BGA: A New Concept in Thermoplastic Packaging', IEEE 29th International Electrical Manufacturing Technology Symposium, July 2004, pp. 345–354.

[28] R. J. Ross, 'LCP Injection Molded Packages – Keys to JEDEC 1 Performance', IEEE 54th Electrical Computational Technology Conference, June 2004, pp. 1807–1811.

[29] D. C. Thompson, O. Tantot, H. Jallageas, G. E. Ponchak, M. M. Tentzeris and J. Papapolymerou, 'Characterization of Liquid Crystal Polymer (LCP) Material and Transmission Lines on LCP Substrates from 30 to 110 GHz', *IEEE Transactions on Microwave Theory and Techniques*, **52**, April 2004, 1343–1352.

[30] C. Khoo, B. Brox, R. Norrhede and F. Maurer, 'Effect of Copper Lamination on the Rheological and Copper Adhesion Properties of a Thermotropic Liquid Crystalline Polymer Used in PCB Applications', *IEEE Transactions on Component Packaging and Manufacturing Technology*, **20**, July 1997, 219–226.

[31] T. Suga, A. Takahashi, K. Saijo and S. Oosawa, 'New Fabrication Technology of Polymer/Metal Lamination and Its Application in Electronic Packaging', IEEE 1st International Polymers and Adhesives in Microelectronics and Photonics Conference, October 2001, pp. 29–34.

[32] X. Wang, L. Lu and C. Liu, 'Micromachining Techniques for Liquid Crystal Polymer', 14th IEEE International MEMS Conference, January 2001, pp. 21–25.

[33] K. Brownlee, S. Bhattacharya, K. Shinotani, C. P. Wong and R. Tummala, 'Liquid Crystal Polymers (LCP) for High Performance SOP Applications', 8th International Advanced Packaging Materials Symposium, March 2002, pp. 249–253.

[34] J. Kivilahti, J. Liu, J. E. Morris, T. Suga and C. P. Wong, 'Panel-Size Component Integration (PCI) with Molded Liquid Crystal Polymer (LCP) Substrates', IEEE Electronic Components and Technology Conference, May 2002, pp. 955–961.

[35] T. Suga, A. Takahashi, M. Howlander, K. Saijo and S. Oosawa, 'A Lamination Technique of LCP/Cu for Electronic Packaging', 2nd International IEEE Polymers and Adhesives in Microelectronics and Photonics Conference, June 2002, pp. 177–182.

[36] T. Zhang, W. Johnson, B. Farrell and M. St Lawrence, 'The Processing and Assembly of Liquid Crystalline Polymer Printed Circuits', *Proceedings of the International Society for Optical Engineering*, 2002, 1–9.

[37] L. Chen, M. Crnic, L. Zonghe and J. Liu, 'Process Development and Adhesion Behavior of Electroless Copper on Liquid Crystal Polymer (LCP) for Electronic Packaging Application', *IEEE Transactions on Electronics Packaging Manufacture*, **25**, October 2002, 273–278.

[38] Modern Machine Shop Online, http://www.mmsonline.com/articles/ 030107.html

[39] PMTEC LCP Materials Symposium, Huntsville, Alabama, 29 October 2002.

[40] H. Kanno, H. Ogura and K. Takahashi, 'Surface Mountable Liquid Crystal Polymer Package with Vertical Via Transition Compensating Wire Inductance up to V–Band', *IEEE MTT-S International Microwave Symposium Digest*, **2**, June 2003, 1159–1162.

[41] M. F. Davis, S.-W. Yoon, S. Pinel, K. Lim and J. Laskar, 'Liquid Crystal Polymer-Based Integrated Passive Development for RF Applications', *IEEE MTT-S International Microwave Symposium Digest*, **2**, June 2003, pp. 1155–1158.

[42] G. Zou, H. Gronqvist, P. Starski and J. Liu, 'High Frequency Characteristics of Liquid Crystal Polymer for System in a Package Application', IEEE 8th International Advanced Packaging Materials Symposium, March 2002, pp. 337–341.

[43] G. Zou, H. Gronqvist, J. P. Starski and J. Liu, 'Characterization of Liquid Crystal Polymer for High Frequency System-in-a-Package Applications', *IEEE Transactions on Advanced Packaging*, **25**, November 2002, 503–508.

[44] D. Thompson, P. Kirby, J. Papapolymerou and M. M. Tentzeris, 'W-Band Characterization of Finite Ground Coplanar Transmission Lines on Liquid Crystal Polymer (LCP) Substrates', IEEE Electronic Components Technology Conference, May 2003, pp. 1652–1655.

[45] Z. Wei and A. Pham, 'Liquid Crystal Polymer (LCP) for Microwave/Millimetre Wave Multi-layer Packaging', *IEEE MTT-S International Microwave Symposium Digest*, **3**, June 2003, 2273–2276.

[46] C. G. L. Khoo, B. Brox, R. Norrhede and F. H. J. Maurer, 'Effect of Copper Lamination on the Rheological and Copper Adhesion Properties of a Thermotropic Liquid Crystalline Polymer Used in PCB Applications', *IEEE Transactions on Component Packaging and Manufacturing Technology*, **20**, 1997, 219–226.

[47] K. Jayaraj and B. Farrell, 'Liquid Crystal Polymers and Their Role in Electronic Packaging', *Advancing Microelectronics*, 1998, 15–18.

[48] Kapton® polymide film, http://www.dupont.com/

[49] L. M. Franca-Neto, R. E. Bishop and B. A. Bloechel, '64 GHz and 100 GHz VCOs in 90 nm CMOS Using Optimum Pumping Method', IEEE International Solid-State Circuits Conference Digest Technical Papers, February 2004, pp. 444–445.

[50] R.-C. Liu, H.-Y. Chang, C.-H. Wang and H. Wang, 'A 63 GHz VCO Using a Standard 0.25 um CMOS Process', IEEE International Solid-State Circuits Conference Digest Technical Papers, February 2004, pp. 446–447.

[51] M. Tiebout, H.-D. Wohlmuth and W. Simbürger, 'A 1 V 51 GHz fully integrated VCO in 0.12 μm CMOS', *IEEE International Solid-State Circuits Conference Digest*, February 2002, 238–239.

[52] K.-W. Yu, Y.-L. Lu, D.-C. Chang, V. Liang and M. F. Chang, 'K-Band Low-Noise Amplifiers Using 0.18 μm CMOS Technology', *IEEE Microwave Wireless Component Letters*, **14**(3), March 2004, 106–108.

[53] L. M. Franca-Neto, B. A. Bloechel and K. Soumyanath, '17 GHz and 24 GHz LNA Designs Based on Extended-S-Parameter with Microstrip-on-Die in 0.18 μm Logic CMOS Technology', *Proceedings of European Solid-State Circuits Conference*, September 2003, 149–152.

[54] X. Guan and A. Hajimiri, 'A 24 GHz CMOS Front-End', *Proceedings of European Solid-State Circuits Conference*, September 2002, 155–158.

[55] M. Madihian, H. Fujii, H.Yoshida, H. Suzuki and T.Yamazaki, 'A 1–10 GHz, 0.18 μm-CMOS Chipset for Multi-mode Wireless Applications', *IEEE MTT-S International Microwave Symposium Digest*, June 2001, 1865–1868.

[56] I. Bahl and P. Bhartia, *'Microwave Solid State Circuit Design'*, 2nd edition, John Wiley & Sons, Inc., Hoboken, New Jersey, 2003.

[57] Chinh H. Doan, Sohrab Emami, Ali M. Niknejad and Robert W. Brodersen, 'Millimetre-Wave CMOS Design', IEEE Journal of Solid-State Circuits, **40**(1), January 2005, 144–156.

[58] T. C. Edwards and M. B. Steer, *'Foundations of Interconnect and Microstrip Design'*, 3rd edition, John Wiley & Sons, Inc., New York, 2000.

[59] B. Kleveland, C. H. Diaz, D. Vook, L. Madden, T. H. Lee and S. S. Wong, 'Exploiting CMOS Reverse Interconnect Scaling in Multigigahertz Amplifier and Oscillator Design', *IEEE Journal of Solid-State Circuits*, **36**(10), October 2001, 1480–1488.

[60] K. Nishikawa *et al.*, 'Miniaturized Millimetre-Wave Masterslice 3-D MMIC Amplifier and Mixer', *IEEE MTTT Transactions*, **47**(1), September 1999, 1856–1862.

[61] Ching-Kuang C. Tzuang, Chih-Chiang Chen and Wen-Yi Chien, 'LC-Free CMOS Oscillator Employing Two-Dimensional Transmission Line', Jointly with the 17th European Frequency and Time Forum Proceedings of the 2003 IEEE International Frequency Control Symposium and PDA Exhibition, 2003, pp. 487–489.

[62] Sen Wang, and Ching-Kuang C. Tzuang, 'Compacted Ka-Band CMOS Rat-Race Hybrid Using Synthesized Transmission Lines', *International Microwave Symposium Digest*, 2007, 1023–1026.

[63] D. R. Smith, W. J. Padilla, D. C. Vier, S. C. Nemat-Nasser and S. Schultz, 'Composite Medium with Simultaneously Negative Permeability and Permittivity', *Physical Review Letters*, **84**(18), May 2000, 4184–4187.

[64] B.-I. Wu, W. Wang, J. Pacheco, X. Chen, T. Grzegorczyk and J. A. Kong, 'A Study of Using Metamaterials as Antenna Substrate to Enhance Gain', *Progress in Electromagnetics Research*, **51**, 2005, 295–328.

[65] Daniela Staiculescu, Nathan Bushyager and Manos Tentzeris, 'Microwave/ Millimetre Wave Metamaterial Development Using the Design of Experiments Technique', IEEE Applied Computational Electromagnetics Conference, April 2005, pp. 417–420.

[66] E. Ozbay, K. Aydin, E. Cubukcu and M. Bayindir, 'Transmission and Reflection Properties of Composite Double Negative Metamaterials in Free Space', *IEEE Transactions on Antennas and Propagation*, **51**(10), October 2003, 2592–2595.

[67] R. A. Shelby, D. R. Smith and S. Schultz, 'Experimental Verification of a Negative Index of Refraction', *Science*, **292**, April 2001, 77–79.

[68] G. V. Eleftheriades, A. K. Iyer and P. C. Kremer, 'Planar Negative Refractive Index Media Using Periodically L-C Loaded Transmission Lines', *IEEE Transactions on Microwave Theory and Techniques*, **50**(12), December 2002, 2702–2712.

[69] A. Grbic and G. V. Eleftheriades, 'Experimental Verification of Backward-Wave Radiation from a Negative Refractive Index Material', *Journal of Applied Physics*, **92**(10), November 2002, 5930–5935.

[70] C. Caloz, T. Itoh, 'Novel Microwave Devices and Structures Based on the Transmission Line Approach of Meta-Materials', *IEEE MTT-S International Microwave Symposium Digest*, **1**, June 2003, 195–198, ISSN: 0149–645X.

[71] Carsten Metz, 'Phased Array Metamaterial Antenna System', US Patent Issued on 25 October 2005.

[72] M. Lancastert, Z. Wut, Y. Huangt, T. S. M. Macleant, X. Zhout, C. Gought and U. McN. Alfords, 'Superconducting Antennas', *Superconductor Science Technology*, **5**, 1992, 277–279.

[73] R. C. Hansen, 'Superconducting Antennas', *IEEE Transactions on Aerospace and Electronic Systems*, **26**(2), March 1990, 345–355.

[74] O. G. Vendik, I. B. Vendik and D. V. Kholodniak, Applications of High-Temperature Superconductors in Microwave Integrated Circuits', *Materials Physica and Mechanics Journal*, **2**(1), 2000, 15–24.

[75] Zhi-Yuan Shen, *'High-Temperature Superconducting Microwave Circuits'*, Artech House, Boston and London, 1994.

[76] M. J. Lancaster, 'Passive Microwave Device Applications of High-Temperature Superconductors' Cambridge University Press, 1997.

[77] O. G. Vendik, I. B. Vendik and D. V. Kholodniak, 'Applications of High-Temperature Superconductors in Microwave Integrated Circuits', Advanced Study Centre Company Ltd, 2000.

[78] I. B. Vendik, O. G. Vendik, S. S. Gevorgian, M. F. Sitnikova and E. Olsson, 'A CAD model for microstrips on R-cut sapphire substrates,' *Microwave Millimeter-Wave Computer-Aided Eng.*, 4, October 1994, 374–383.

[79] E. K. Hollmann, O. G. Vendik, A. G. Zaitsev and B. T. Melekh, *Superconductor Science Technology*, 7, 1994, 609.

[80] C. S. Gorter and H. Casimir, *Physica Zeitschrift*, 35, 1934, 963.

[81] K. Huang, 'High Temperature Superconducting Microwave Devices', PhDThesis, University of Oxford, 2000.

[82] G. W. Hanson, 'Fundamental Transmitting Properties of Carbon Nanotube Antennas', IEEE Transactions on Antennas and Propagation, 53(11), November 2005, 3426–3435.

[83] Nikolaus Fichtner and Peter Russer, 'On the Possibility of Nanowire Antennas', *Proceedings of the 36th European Microwave Conference*, 2006, 870–873.

[84] Yue Wang, Qun Wu, Wei Shi, Xunjun He *et al.*, 'Radiation Properties of Carbon Nanotubes Antenna at Terahertz/Infrared Range', *International Journal of Infrared Milliwaves*, 29(1) 35–72, January 2008, Springer New York, ISSN: 0195-9271.

10

High-Speed Wireless Applications

Antennas are one of the key building blocks for wireless communications networks. There are innumerable possible applications for millimetre wave antennas, but this chapter will give some examples of V-band and E-band antenna applications in Sections 10.1 and 10.2. Distributed antenna systems and wireless mesh networks are discussed at the end of the chapter.

10.1 V-Band Antenna Applications

The 60 GHz band offers an ample, license-free bandwidth. In the United States, the range from 57 to 64 GHz is available, while in Japan, 59 to 66 GHz is available (see Chapter 1). With 7 GHz of bandwidth, there are many high data rate applications that can be envisage. The 60 GHz radio is suitable for high data rate and short distance applications, as there is less interference with other wireless standards than the UWB [1]. Therefore it can be used for indoor applications such as audio/video transmission, desktop connection and for the support of portable devices. These applications can be divided into the following categories:

- Personal area network
- High-definition video streaming
- Point-to-multipoint links
- Video broadcasting
- Intervehicle communication
- Multigigabit file transmission

In each category, there are various situations that can arise based on:

- whether they are used in a residential area or office
- distance between the transmitters and receivers
- line-of-sight (LOS) or non-line-of-sight (NLOS) connections
- position of the transceivers
- mobility of the devices

Millimetre Wave Antennas for Gigabit Wireless Communications   Kao-Cheng Huang and David J. Edwards
© 2008 John Wiley & Sons, Ltd

10.1.1 Wireless Personal Area Networks (WPANs)

The 60 GHz band is ideally suited for personal area network (PAN) applications [2]. A 60 GHz link could be used to replace various cables used today in the office or home, including gigabit Ethernet (1000 Mbps), USB 2.0 (480 Mbps) or IEEE 1394 (~800 Mbps). Currently, the data rates of these connections have precluded wireless links, since they require so much bandwidth. The 60 GHz band is providing promising wireless technologies for these applications. The intended range of WPANs is 10 m or less, which covers the size of most offices, medium-size conference rooms and rooms in the home. Wireless PANs could interconnect various electronic devices, including laptops, cameras, PDAs and monitors. Applications include wireless displays, a wireless docking station and wireless streaming of data from one device to an other [3]. Streaming data from one device to another will benefit from the high data rates achievable at 60 GHz. For example, a *Blu-ray Disc* with 50 gigabytes of memory will take more than two hours to download its contents over a 54 Mbps WiFi connection, but ~50 seconds to download over a 1 Gb/s at 60 GHz.

10.1.2 Wireless HDMI

The high-definition multimedia interface (HDMI) is evolving to be the standard interface for high-definition TVs (see Figure 10.1). This cable provides both video and audio information. Depending on the resolution of the display, the data rates required for an uncompressed HDMI signal can be substantial. The key advantage of 60 GHz is the ability to provide wireless, secure and uncompressed high-definition video distribution [4]. Wireless networks allow the display to be located far from the information source (DVD player, cable box, etc.). This obviates the need for bulky wires from the "picture-frame" display on the wall to the DVD player in the cabinet. Security is provided at 60 GHz due to the atmospheric and material properties at this frequency; over long ranges, there is significant signal loss due to oxygen absorption, and there is also significant attenuation through walls. These two facts prevent the HDTV signal from leaking into adjacent rooms and residences. This is a definite benefit when it comes to content providers, who desire to limit distribution of the contents to the legitimate purchaser of their services. Uncompressed signalling retains the image quality in the link, which is of particular concern if lossy compression is being considered. Additionally, avoiding compression avoids the need to pay royalties for the compression algorithm. The data rates for HDMI depend on the resolution of the display and the use of the interlacing or progressive scan. High-definition video streaming includes uncompressed video streaming for residential use. In this application an uncompressed

mm wave camcorder

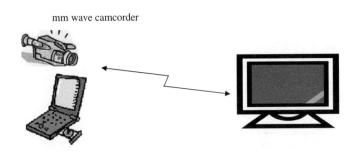

Figure 10.1   Wireless high-definition applications

video/audio stream is sent from a DVD player to an HDTV. A typical distance between them would be 5 to 10 m with either a LOS or NLOS connection [5]. The high-definition streams can also come from portable devices such as a laptop computer, personal data assistant (PDA) or portable media player (PMP), placed somewhere in the same room with an HDTV. In this setting, the coverage range might be 3 to 5 m with either a LOS or NLOS connection. NLOS may produce the best results because the direct propagation path may be temporarily blocked by human bodies or objects. Uncompressed video streaming can also be used for a laptop-to-projector connection in a conference room where people can share the same projector and easily connect to the projector without switching cables, as has to happen in the case of cable connection.

The WirelessHD 1.0 specification was launched in January 2008. It is architected and optimised for wireless display connectivity, achieving in its first generation implementation high-speed rates from 2 to 5 Gb/s for the CE, PC and portable device segments. Its core technology projects theoretical data rates as high as 20 Gb/s, permitting it to scale to higher resolutions, colour depth and range. Coexisting with other wireless services, the WirelessHD platform is designed to operate cooperatively with existing, wireline display technologies.

The recent availability of several new technologies makes it possible to achieve the multi-gigabit data rates required for uncompressed video streaming [6]. Such breakthroughs enable low cost, better image quality and higher performance wireless A/V systems. The major characteristics and key technologies include:

- High interoperability supported by major CE device manufacturers
- Uncompressed HD video, audio and data transmission, scalable to future high-definition A/V formats
- High-speed wireless, multigigabit technology in the unlicensed 60 GHz band
- Smart antenna technology to overcome line-of-sight constraints of 60 GHz
- Secure communications
- Device control for simple operation of consumer electronic products
- Error protection, framing and timing control techniques for a quality consumer experience

10.1.3 Point-to-Point 60 GHz Links

Point-to-point links are used today for telecommunications backhauls [7]. They employ high-gain antennas to increase the range of the link. The 60 GHz band is being used in the marketplace today for such links, and the corresponding chips are implemented in III–V technologies. Silicon offers the best promise to reduce the cost of these systems, though the bulk of the system cost is still installation fees, requiring someone to install the equipment on top of a pole. In Figure 10.2, ad hoc information distribution with a point-to-point network extension makes advertisement distribution or contents downloading services easy, and allows immediate construction, for example in an exhibition site.

10.1.4 Broadcasting a Video Signal Transmission System in a Sports Stadium

Using the 60 GHz technology on an HDTV camera system, images and audio signals can be transferred from a video camera to a monitoring or recording facility. Such a system is used

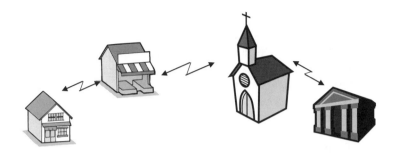

Figure 10.2   Ad hoc information distribution with a point-to-point network

in cases when signal transfer through the wire is challenging, such as in sports broadcasting or security systems, etc.

10.1.5 Intervehicle Communication System

In Figure 10.3, group cooperative driving requires that vehicle control data (regarding speed, acceleration, steering situation, etc.) for each vehicle is transmitted between the user's vehicle and the vehicles operating nearby. Using this technique, controlled cruising and assisted merging/diverging are accomplished. To support these functions, a vehicle-to-vehicle intervehicle communications (IVC) system is of course required.

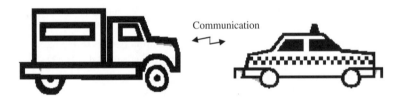

Figure 10.3   Intervehicle communication system

10.1.6 Multigigabit File Transmission

Enhancing the performance of the broadband Internet and the performance of computer and storage systems requires high-bandwidth networks to interconnect to these systems. The 60 GHz technology has already marketed high-speed network interface products, such as the 10 G Ethernet, that accommodates high-speed data transfer at 6.4 Gb/s per signal line to increase network bandwidth [5]. Figure 10.4 shows an application of the 60 GHz technology which was designed to support high-volume upload/download processes. It also employs an intuitive, graphic interface presenting a side-by-side client–host file system display above a convenient session list, and saves commonly transferred files, including subfolders and contents, for repeated future use.

10.1.7 Current Developments

There are numerous possibilities for millimetre wave antennas to support a wide range of applications. The spectrum at 60 GHz is now open to consumer electronics and mass market

Figure 10.4 Large scale data transmission

applications. Several equipment suppliers have developed transceiver devices for use in the 60 GHz band. Matsushita, Panasonic, IBM, Philips, Korea University, Terabeam, Telaxis and France Telecom Company are a few of the primary suppliers of these devices. A number of millimetre wave systems are listed below, along with short technical descriptions of their specifications.

On/Off Keying System
The on/off keying (OOK) modulation technique was proposed for portable applications in Reference [8]. The approach enables the simplest radio architecture to realise the lowest cost and fastest time to market. It is suitable for Kiosk file downloading and portable point-to-point applications. The typical range for file downloading/uploading is 1 m. The advantages of OOK modulation are:

- Very simple, with a sufficient rate for portable devices (above 1Gpbs)
- Very low power consumption compared with other modulations
- Works well on additive white Gaussian noise (AWGN)/Kiosk/residential LOS channels
- Very suitable for portable point-to-point applications
- Simple coexistence with BPSK/QPSK (binary/quadrature phase-shift keying) and other signal devices
- Phase noise of the local oscillator has no effects on the detection performance

Minimum Shift Keying Systems
IBM Research proposes multigigabit per second communications at 60 GHz based on MSK modulation [9]. A minimum shift keying (MSK)-based system for multigigabit wireless communication at 60 GHz presents significant advantages, e.g. lower complexity and power consumption in a directional channel. The typical range is about 1 m and the bit rate is 2 Gb/s. MSK is an excellent choice due to its bandwidth efficiency and ease of implementation. This type of modulation is attractive for high-volume, low-power and low-cost point-and-shoot applications in multigigabit wireless communications. The following is the technical specification of a 60 GHz product:

- Transmit power: 7.0–7.5 dBm
- EIRP: +14.5 dBm
- $E_s/N_0 \sim 10$ dB

The Nokia Metro Hopper radio is a wireless access link using a 57.2–58.2 GHz frequency range for reliable transmission [10]. Part of the Nokia Metro Site capacity solution, this radio provides access to Nokia Metro Site base stations in dense microcellular networks. Compact and unnoticeable, the Nokia Metro Hopper radio enables a variety of new site locations to be used and can eliminate the costly delays associated with traditional cellular transmission. The main benefits provided by the Nokia Metro Hopper are:

- Fast base station access eliminates costly delays caused by frequency licence applications or leased line availability.
- Unique 58 GHz frequency band enables high-density networks.
- Reliable microwave access improves network quality to help reduce subscriber dissatisfaction.
- Lightweight and compact design allows innovative urban locations to simplify site acquisition and lower costs.

Listed below are the typical specifications:

- Transmission capacity: 4×2 Mbit/s
- Antenna type: flat panel antenna with gain of 34 dB
- Receiver threshold level BER of 10^6 with less than 71 dBm input signal level
- Transmitter: output power of 5 dBm

Orthogonal Frequency Division Multiplexing

Orthogonal frequency division multiplexing (OFDM)-based PHY is proposed to allow data rates from 335 Mbps to 3 Gb/s using 60 GHz transmission [11]. The technical advantages of OFDM are that it is inherently robust against any type of channel fading, providing high spectrum efficiency and allowing high data rates to be reached. OFDM is a future-proof technology because:

- Is a widely used technology (WiFi, WiMax, DAB, DVB, ECMA UWB)
- Large scope of possible applications: from point-to-point data transfer to cell mode coverage
- Compatible with advanced techniques: beamforming, MIMO STBC, etc

Key features of OFDM are [11]:

- Data rates from 335 Mbps to 3 Gb/s for applications such as video streaming, file transfer, home network distribution or in-vehicle media supply
- Efficient channellisation adapted to worldwide regulation
- OFDM-based system providing high spectrum efficiency
- Scalable parameters for increased robustness
- Low-power and cost-effective implementation

Single-Carrier Block Transmission

Single-carrier block transmission (SCBT) is the bridge between OFDM and an SC (single carrier) [12]. SCBT is a form of SC with a low peak-to-average power ratio (PAPR), and can use weak codes or is unencoded. SCBT is also a form of OFDM (with Fourier spreading) with frequency domain equalisation in MP, and has the flexibility to use a time-domain equalizer or no equaliser at all (in AWGN). Table 10.1 shows the advantages of SCBT.

Table 10.1   Single-carrier block transmission [12]

Advantages of SCBT (compared to OFDM)	
Low PAPR	• At 60 GHz, dynamic range of PAs is limited. • An OFDM system has to back-off considerably more than SCBT. • If multiple PAs are used, power back-off is not necessary, but power consumption increases with PAPR.
Better performance with high rate or weak codes	• For high rate (multi-Gb/s) modes, high rate and/or weak codes must be used. Implementing a complex decoder at multi-Gb/s is a problem. • OFDM gets diversity from the code but it does not perform well with high rates or weak codes. • SCBT gets diversity from spreading data over the whole band.
Fewer bits in the ADC block	• SCBT requires fewer bits in the ADC, when the channel has a low number of multipaths.
For flexibility in receiver design	• Both frequency-domain and time-domain receivers can be used. • It is a bridge between OFDM and a single carrier.

10.2 E-Band Antenna Applications

The E-band products address the requirements of carriers, enterprises and cable, and government and Internet service providers building cost-efficient, wireless multigigabit IP networks. The E-band technology provides solutions for interconnection and the backhaul of 4G, WiMAX, mobile networks, distributed antenna systems and remote radio heads (RRH), gigabit Ethernet access network connections, last mile access, fibre back-up and network extension applications [13, 14].

10.2.1 Private Networks/Enterprise LAN Extensions

In Figure 10.5, network administrators operating LAN networks in campus environments often face the challenge of establishing a private, high-speed network connection between their different buildings [15]. This could be for simple Ethernet LAN extensions, for offsite back-up of files or for transfer of customer data, billing, images or other large data files. Instead of laying a fibre-optic cable or leasing an expensive lower-speed network connection,

Figure 10.5   Private networks/enterprise LAN extensions

the operator can use an E-band radio to establish network connectivity between the remote locations.

10.2.2 Fibre Extensions

Fibre is an ideal medium for high data rate transmission. However, it has high cost and takes several months to trench and commission, so fibre has difficulty in getting across rivers and railway lines. Figure 10.6 shows that E-band technology is ideally suited to complement fibre; in that it permits fibre-like speeds of 10 Gb/s in a quick and easy-to-deploy wireless configuration.

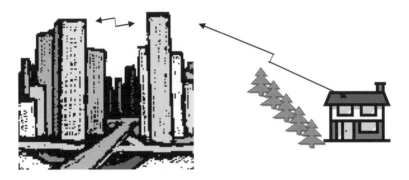

Figure 10.6   Fibre extensions between buildings

10.2.3 Fibre Back-up/Diversity Connections

Some buildings already come with high-speed connections. For those applications with the most sensitive data needs or requiring thorough disaster preparedness, it is necessary to back up existing connections with a technology offering different failure mechanisms and ideally different entry/exit points to the building. The Federal Government in the United States, for example, requires all its federal buildings to have diversity backed-up telecommunication systems. E-band is an ideal back-up to high-speed optical fibre, as it operates with optical fibre-like speeds of 10 Gb/s, and being wireless will most likely enter the building via a roof or high window. This makes the wireless link impervious to the most common causes of fibre outages, such as flooding, construction work, earthquakes and other ground-based disturbances.

10.2.4 Military Communications and Surveillance Systems

Wireless is the preferred choice for military communications and surveillance in combat scenarios and hostile battlefield environments. When properly commissioned, wireless can be quick to install, highly portable and inherently secure. This 70/80 GHz technology offers the ability to wirelessly transmit an unprecedented 10 Gb/s, significantly more data than current secure systems. This makes possible the transmission of real-time high-definition battlefield images or large data file transfer in a quasi-mobile hostile environment. Alternatively, multiple uncompressed high-definition HDTV images can be transmitted and interconnected, allowing real-time monitoring of multiple security images or viewing of secure locations at the same time.

10.2.5 Secure Applications

E-band systems offer huge military and government potential. Very high data rates, portability and ease of deployment make E-band communications ideal for homeland defence and other security purposes. Rapidly deployable, high-definition video links can be used for high data rate for event security such as facial mapping for recognition [16]. Another is high data rate battlefield data distribution from satellites. E-band systems excel in military applications since the millimetre wave properties allow the radio to be classified as low-probability-of-detect/low-probability-of-intercept (LPD/LPI). For any point-to-point technology, antennas generally have to be placed on high towers or masts, so a clear line-of-sight can be achieved. By necessity, this means that few buildings or possible interception points can be close to the radio path. However, the ether between any two antennas is never secure, and a determined hacker with a basic understanding of the particular radio system architecture will be able to intercept part of the broad transmission radio signal. E-band systems, however, have a frequency much higher than conventional wireless systems, resulting in antennas that have a much greater focusing ability. A typical application with a 0.6 m dish antenna has a very narrow 0.3° beamwidth, meaning that transmitted signals propagate as highly focused and directional pencil beams from the antenna. Transmitted signals do not form broad diverging beams that illuminate large areas as at lower frequencies. For the 0.6 m E-band scenario, the beam will spread in a cone shape over 30°. Thus, an E-band product's narrow pencil beams are inherently much harder to locate and are far more difficult to intercept.

Table 10.2 shows the comparison between all the high data rate transmission technologies. The 70/80 GHz technology gives the best performance compared with other technologies.

Table 10.2 Comparison of the performance of all the available high data rate transmission technologies [17]

	70/80 GHz	60 GHz radio	Free space optics	Buried fibre
Data rates	1 Gb/s	1 Gb/s	1 Gb/s	Virtually unlimited
Typical link distances (99.999 % availability)	1.6 km	366 m	183 m	Virtually unlimited
Relative product complexity	Low	Low	High	Low
Relative cost of installation and ownership	Low	Low	Low	High
Installation time	Hours	Hours	Hours	Months
Regulatory protection	Yes	No	No	Yes

In conclusion, E-band wireless communications is a new technology that allows gigabit per second (Gb/s) data rates to be transmitted with very high weather tolerances over distances of a mile or more. Characterised as LPD/LPI, it is a perfect technology to satisfy hostile territory battlefield situations, where there is a need for high-security, high-speed, point-to-point and non-wire-line communications. A novel licensing structure coupled with an ability to deploy links quickly permits a rapid response to time-critical security applications. In the modern world, essentially all communications, from telephone calls to personal e-mails to e-commerce transactions, are carried by digital networks. The ubiquity and ease of access to digital networks coupled with the sensitivity of much of the transmitted data, means that security is now a prerequisite for these systems. Wireless communication systems traditionally have not provided any level of security or privacy. In fact, the early value of wireless was its ability to be widely detected and decoded (e.g. terrestrial TV and radio broadcasts). However, the growth of data communications for Internet and e-commerce services has forced system designers to start considering security as a primary system requirement.

10.3 Distributed Antenna Systems

Distributed antenna systems (DAS) use one base station and a series of hubs to distribute the radio signal through multiple remote antennas. A distributed antenna system has several spatially separated antenna nodes that are connected to a common source via a transport medium to cover a geographic area or structure. DAS antenna elevations are generally at or below the clutter level, and node installations are compact.

As illustrated in Figure 10.7, the idea is to split the transmitted power among several antenna elements, separated in space so as to provide coverage over the same area as a single antenna but with reduced total power and improved reliability. A single antenna radiating at high power (as shown in the grey area) is replaced by a group of low-power antennas to cover the same area, as shown inside the four circles. This idea was published in 1987 [18].

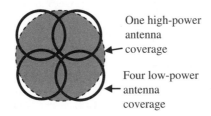

One high-power
antenna
← coverage

Four low-power
← antenna
coverage

Figure 10.7   Coverage of a conventional antenna and distributed antennas

The advantage of this green concept is that less power is wasted in overcoming penetration and shadowing losses, and because a line-of-sight channel is present more frequently, there are reduced fade depths and reduced delay spread.

A distributed antenna system can be implemented using RF splitters and feeders, or a repeater. Amplifiers can be included to overcome the feeder losses. If equalisation is applied to the system, it may be desirable to introduce delays between the antenna elements. This artificially increases delay spread in areas of overlap coverage and so improves wireless link quality due to time diversity of signal arrival.

If a given area such as a campus, airport or town centre is covered by many distributed antenna elements rather than a single antenna, then the total radiated power is reduced by approximately a factor of $N^{1-n/2}$ and the power per antenna is reduced by a factor of $N^{n/2}$, where a simple power law path loss model with a path loss exponent n is assumed. As an alternative, the total area covered could be extended for a given limit of effective radiated power, which may be important to ensure compliance with safety limits on radiation exposure of the human body.

Some distributed antenna systems use large coaxial cables to distribute the signal through a venue: typically, coaxial cable is used in the building risers, with coaxial cable diverging from this backbone, running horizontally and feeding antennas at the run's endpoint. These systems are considered to be passive because they do not use managed electronics in the distribution network.

A passive distributed antenna system is expensive to install (large coaxial cable requires special installation labour and installation time can be lengthy). In addition, the signal in a passive system attenuates over longer cable runs, leading to inconsistent signal strength. This makes system design/engineering difficult as each antenna's coverage area will vary, depending on the length of coax to which it is attached. This leads to another issue: with a passive distributed antenna system, the whole system may require re-engineering if additional capacity is added or if additional coverage areas are required. Finally, passive systems are difficult or impossible to manage: if an antenna malfunctions, only user complaints will alert the operator. Typically, passive distributed antenna systems are deployed in smaller facilities where the length of a coaxial cable run is less of an issue.

Passive distributed antenna systems, like most technologies, have their problems. The three primary disadvantages to distributed antenna system solutions are cost, carrier costs and a lack of standards:

1. Costs for in building coverage can exceed a dollar or more per square foot; this adds up quickly in a large building. Costs to supplement outdoor coverage must be determined individually, since there are many variables. One way to economise in a distributed antenna system (DAS) installation is to equip the building during the building process, which means a network design needs to be developed in advance and the necessary trunking put in place for the routing of DAS cables both horizontally and vertically, as is implemented for plumbing, electricity and Ethernet. This early design in conjunction with the architect should also include determining where the head end should be placed and selecting cable routes to the roof for antennas.

2. Cooperative wireless carriers. In order to extend or enhance wireless service in a domain, the owner needs active support from its service providers. Carriers are responsible for compliance with national regulations such as avoiding interference between systems, so if their signals are to be re-broadcast by a third party, the carriers must ensure such compliance. Frequently however, carriers are not interested in getting involved in implementation unless they anticipate a significant increase in the traffic and revenue. They also may be reluctant if they believe competition will be too great in a multiple-provider situation. If good coverage is required from multiple providers, consideration needs to be given to a "neutral host" type of system. Typically, those that implement these systems have influence to bring multiple service providers together. The role may also include negotiating revenue-sharing agreements with these carriers to minimise overall maintenance costs.

3. Standards. Here, neither the IEEE nor the ITU has developed a standard for distributed antenna systems. This unfortunate situation is similar to that of networking, before Ethernet became the accepted standard: with many competing proprietary technologies. It is therefore essential that those who select the operating system understand the pros and cons of the various manufacturers' products and the associated methodologies of the designs. Unbiased third parties (such as independent consultants) can assist by helping assess the wireless needs of the client, and then to develop budgetary estimates to give an idea of the range of costs likely to be incurred.

Active distributed antenna systems utilise active hubs, fibre optics, active radio access units (RAUs) and antennas to distribute the signal. These systems sustain high signal strength at every antenna location, independent of the distance from the central transceiver. Some active distributed antenna systems extend over several miles, and many systems support optimum signal strength throughout the interiors of buildings and other facilities measuring millions of square metres. In addition, active distributed antenna systems offer end-to-end monitoring and management down to the individual antenna, ensuring that outages will quickly be reported and repaired.

Active DAS systems address all of the problems that are met when deploying pico cell systems:

1. Costs. Active distributed antenna systems are much smaller, cost less and are less intrusive than pico cells aesthetically. In addition, active distributed antenna systems use standard cabling to connect hubs to RAUs and remote antennas, so this expedites installation and reduces costs when compared to passive distributed antenna systems.
2. Terminal density. With an active DAS, there is no limit to the number of antennas. Each antenna is an extension of the central radio source (all antennas in the system broadcast all channels from the radio source), so there is no need to design the system at each individual antenna level to meet the needs in hot spot areas.
3. Interference. Because the system operates off of a single radio source, there is no problem with interference among antennas.
4. Longer term savings. Distributed antenna systems use low cost RAUs and antennas to support high-capacity areas, users need to invest less capital in expensive access points that may be infrequently used.
5. Multi Carrier Capability. Unlike pico cells, both active and passive distributed antenna systems can support multiple carriers in a single-antenna system, making it a relatively easy matter to upgrade centralised or hub electronics so they can support new carriers or services.
6. Capacity. In an active distributed antenna system, capacity issues are addressed at the central base station. If additional capacity is required, the number of radios at the central source is simply increased. This is in contrast to pico cells, where it may be necessary to add more radio resources at every pico-cell to support extra users, which is costly.
7. Interfaces. Major communications carriers throughout the world have installed large numbers of distributed antenna systems, and therefore the interfaces to base stations and the network are well known. Active distributed antenna systems are the in-building systems chosen to provide coverage inside most major cellular company headquarters.

There are still several issues that remain to be addressed, concerning indoor distributed antenna systems. The emergence of high-speed data services and the move towards providing cellular functions in laptops will mean indoor cellular coverage becomes more important.

10.4 Wireless Mesh Networks

There are two ways at present to obtain relatively high-speed wireless access to the backbone network. This may be achieved either via cellular or via WLAN. They have different performances in terms of quality of service, speed and range, and also in economics. Communications performance is also different; cellular is a connection orientated system biased towards real-time voice, and WLAN for packet based non-real-time data. The concept of having a dual-mode handset for attachment; via a WLAN at close range and via cellular at long range, is very attractive to the user. However, such convergence presents a real challenge to operators who may need to change their business model. The problem of end to end delay in highly congested capacity limited WLANs also presents problems for real time applications, especially in safety sensitive applications.

The vision of mobile communications predicts the future as an integration of all mobile and wireless nodes (e.g. cellular, WLAN, PAN, etc.) with an IP core. This is an ultimate integration of the cellular approach with the WLAN approach, where proprietary interfaces and protocols are effectively dispensed with [19].

In gigabit communications, many approaches could be used to increase capacity and flexibility of wireless systems. Typical examples include directional and smart antennas [20, 21], MIMO systems [22–24], and multiradio/multichannel systems [25].

The wireless mesh network is relatively new, and may enable a new approach to high-speed wireless access.

There are two main types of wireless mesh networks: *infrastructure WMNs* (wireless mesh networks) and *ad hoc WMNs*:

- *Infrastructure WMNs* are often used in conjunction with other communications standards such as community and neighbourhood networks and now can be built using infrastructure meshing. The mesh routers are placed on the roofs of buildings in an area which serves as access points for users inside the buildings and in the street. Usually two types of routers are used, one type for backbone communication and one for user communication. Backbone communication can be established using long-range communication techniques including directional antennas.
- *Ad hoc WMNs*. Client meshing provides peer-to-peer networks between terminals. In this architecture, client nodes support the network by performing routing and configuration functions in addition to providing the communications function. Consequently, a mesh router is not required. The architecture is shown in Figure 10.8 (b). In client WMNs, a packet intended for a particular node in the network hops through multiple nodes to reach its destination. Ad hoc WMNs are usually formed using one type of device. Moreover, increased requirements are placed on end-user devices compared to infrastructure meshing, since, in ad hoc WMNs, the end-user devices must perform additional functions such as routing and self-configuration.

In both networks when strong interference is present, diversity processing alone is insufficient to maintain high quality of service. To resolve this issue, adaptive antenna-array

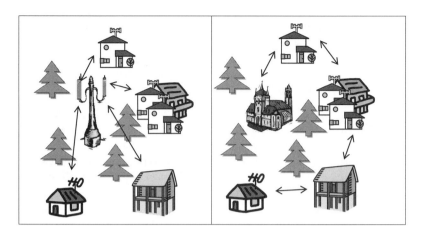

Figure 10.8   (a) Point-to-multipoint network and (b) fixed mesh network

processing is used to shape the antenna beam in order to enhance the desired signals while nulling the interfering signals. Antenna diversity and smart antenna techniques are also applicable to WMNs and other ad hoc networks.

Examples of analysing smart antenna systems for *mobile ad hoc networks* are reported in References [20] and [26]. Due to complexity and cost, a fully adaptive smart antenna system is only used in base stations of cellular networks. Ongoing research and further development is still needed to implement fully adaptive smart antenna systems in a mobile terminal. For WMNs, achieving low cost is a challenge. To this end directional antennas have been actively researched in the area of ad hoc networks.

A mechanically or electronically steerable or switched directional antenna system can be configured to track a terminal signal. By using directional transmission, interference between network nodes can be mitigated and, thus, network capacity can be improved [21, 27]. Directional antennas can also improve power efficiency [28]. However, directional antennas bring problems for the MAC protocol design [29, 30].

Directional-antenna-based MAC schemes can eliminate exposed nodes if the antenna beam is assumed to be perfect. However, due to the directional transmission, more hidden nodes are generated. Therefore, new solutions must be developed to reduce the number of hidden nodes. Ultimately, MAC protocols based on directional antennas face other challenges such as cost, system complexity and the practicality of fast steerable directional antennas.

Most systems support variable bit rate access for a diverse range of mobile devices. This demands a provision of service that is well supported by TCP/IP. The core network is then used to connect the users to the Internet via Internet service providers (ISPs).

Content providers are closely associated with ISPs in some business models. The network classes linked into the core are:

- Personal area network (PAN). This may, for example, use Bluetooth which may support a mesh like architecture. Speed is currently under 1Mb/s (although 100 Mb/s is predicted) and the range is short, e.g. 10 m. No infrastructure is needed, except for an access point to a backbone.

- Mesh. These are presently only very sparsely deployed, mainly in the early user market. Potentially they provide high data rates and, good coverage.

Propagation effects are generally assumed to follow a log-normal fading model. This may be a misinterpretation, but it does provide a very worthwhile physical starting point. The contention is that point-to-multipoint links cope most easily with the log fading of an open environment, but they must use high power to cope with the normal fading environment due to clutter in urban and indoor environments. For the mesh networks it is conjectured that the converse is true, and mesh networks deal well with cluttered environments, but not so well with open environments and longer ranges.

The performance analysis cited above has been carried out in the context of intramesh traffic, with links formed between node-pairs, and with uniform random distribution of nodes and traffic flows. In particular, this model leads to a random distribution of the orientation of radio links and hence a random distribution of interference paths, both of which can help bring about improvements through the use of directional antennas. The case of a mesh system handling extra-mesh services via access points, or intra-mesh traffic via relay nodes to a backbone mesh network is different. When handling such traffic, it is assumed that the links are aligned roughly radially around the nodes (an access point or relay node) and so adjacent mobile nodes have their antennas aligned radially inwards to the fixed node, as illustrated in Figure 10.9. This means that the arrangement of antennas will tend to increase the interference between neighbouring nodes, and so the gains from directional antennas will be negated.

Figure 10.9   Interfering neighbours in a hybrid network

In order to improve the performance of a wireless radio and its control by higher- layer protocols many techniques have been employed. These include more advanced radio technologies, such as reconfigurable radios, frequency agile/cognitive radios [31–33] and even software radios [34]. Many of these radio technologies are still in their early stages of development, they are expected to be the platform for future wireless networks owing to their ability to dynamically control the radio links.

There are two problems associated with the physical layer. Firstly, it is necessary to further improve the transmission rate and the performance of physical layer techniques. New wideband transmission schemes in addition to OFDM or UWB are needed in order to achieve a higher transmission rate in larger area networks. Multiple-antenna systems have been studied for some time. However, their complexity and cost are still too high to be widely accepted for WMNs. A low-cost directional antenna implementation is reported in Reference [35]. Frequency-agile techniques are also at an early stage. Many issues need to be resolved before they can be accepted for commercial use [33].

Secondly, to fully utilise the advanced features provided by the physical layer, higher level protocols, especially MAC protocols, need to be carefully designed. Otherwise, the advantages brought by these physical layer schemes could be compromised. For directional and smart antennas, many MAC protocols have been proposed for ad hoc networks [29, 30]. A MAC protocol for MIMO systems is studied in Reference [23].

Therefore advanced wireless radio technologies all require a revolutionary design in antenna-related technology. For example, when directional antennas are applied to IEEE 802.15.3c networks, a routing protocol needs to take into account the selection of directional antenna sectors. Directional antennas can reduce exposed nodes, but they also generate more hidden nodes. Thus, MAC protocols ought to be redesigned to resolve this issue. The same principle applies to MIMO systems [23]. Therefore, much research work is needed to develop millimetre wave antennas for these future applications.

References

[1] Bridge Wave Communications, 'Gigabit Wireless Applications Using 60 GHz Radios', 2007.
[2] Abbie Mathew, 'Local Area Networking Using Millimetre Waves', NewLANs, Inc., USA, 2005.
[3] Chinh H. Doan, Sohrab Emami, Ali M. Niknejad and Robert W. Brodersen, 'Design of CMOS for 60 GHz Applications', University of California, Berkeley Wireless Research Center, Berkeley, 2005.
[4] S. Li, 'Designing for High Definition Video with Multi-Gigabit Wireless Technologies', Sibeam White Paper, November 2005.
[5] Eino Kivisaari, '60 GHZ MMW Applications', Helsinki University of Technology, Telecommunications Software and Multimedia Laboratory, Finland, 2003.
[6] Nan Guo, Robert C. Qiu, Shaomin S.Mo and Kazuaki Takahashi, '60-GHz Millimeter-Wave Radio: Principle, Technology, and New Results', *EURASIP Journal on Wireless Communications and Networking*, 2007, Article ID 68253, DOI:10.1155/2007/68253.
[7] Dong-Young Kim, '60 GHz SoP Design Using LTCC', Microwave Devices Team, ETRI, 2006.
[8] 'Panasonic PHY and MAC Proposal to IEEE802.15 TG3c CFP', IEEE 802.15 Working Group for Wireless Personal Area Network (WPAN), May 2007.
[9] 'Summary of Link Experiments and MSK-Based Preliminary Proposal for Multi-GB/s Communications at 60 GHz', IEEE 802.15 Working Group for Wireless Personal Area Network (WPAN), March 2007.
[10] Nokia_Metrohopper™ Radio, Nokia leaflet.
[11] France Telecom, 'Proposition of a High Data Rate Wireless System in the 60 GHz Range, Providing Data Rates Ranging from 335 Mbps to 3 Gb/s', IEEE 802.15 Working Group for Wireless Personal Area Network (WPAN), 7 May 2007.

[12] 'SCBT Based 60 GHz PHY Proposa', IEEE 802.15 Working Group for Wireless Personal Area Network (WPAN), May 2007.

[13] Jonathan Wells, 'Multi-Gigabit Wireless Technology at 70 GHz, 80 GHz and 90 GHz', Gigabeam Corporation, 2006.

[14] John Cox, 'E-Band Radio for High-Capacity Links', NetworkWorld.com, 17 February 2006.

[15] 'Millimetre Wave Communication System', Lumera Data Sheet, http://www. lumera.com/home.php

[16] Jonathan Wells, 'New Multi-Gigabit Wireless Systems Satisfy High-Security Rapid Response Applications', Military Embedded Systems, Spring 2006.

[17] Jonathan Wells, White Paper on 'WiMAX Backhaul at 70/80 GHz', Gigabeam Corporation, USA, 2006.

[18] A. A. M. Saleh and R. A. Valenzuela, 'A Statistical Model for Indoor Multipath Propagation, *IEEE Journal of Selected Areas in Communications*, **SAC-5**, February 1987, 128.

[19] Ahmad Atefi, 'Final Report – Study of Efficient Mobile Mesh', OFCOM Document AYR005, January 2006.

[20] R. Ramanathan, 'On the Performance of Ad Hoc Networks with Beamforming Antennas', ACM International Symposium on *'Mobile Ad Hoc Networking and Computing (MOBIHOC)'*, October 2001, pp. 95–105.

[21] A. Spyropoulos and C. S. Raghavendra, Asymptotic Capacity Bounds for Ad Hoc Networks Revisited: The Directional and Smart Antenna Cases', IEEE Global Telecommunications Conference (GLOBECOM), 2003, pp. 1216–1220.

[22] W. Xiang, T. Pratt and X. Wang, 'A Software Radio Testbed for Two-Transmitter Two-Receiver Space Time Coding Wireless LAN', *IEEE Communications Magazine*, **42**(6), 2004, S20–S28.

[23] K. Sundaresan, R. Sivakumar, M. A. Ingram and T.-Y. Chang, 'A Fair Medium Access Control Protocol for Ad Hoc Networks with MIMO Links', IEEE Annual Conference on Computer Communications (INFOCOM), 2004, pp. 2559–2570.

[24] IEEE 802.11 Standard Group Website, <http://www.ieee802.org/11/>

[25] J. So and N. Vaidya, 'Multi-channel MAC for Ad Hoc Networks: Handling Multi-channel Hidden Terminals Using a Single Transceiver', ACM International Symposium on *'Mobile Ad Hoc Networking and Computing (MOBIHOC)'*, May 2004, pp. 222–233.

[26] S. Bellofiore, J. Foutz, R. Govindaradjula, I. Bahceci, C. A. Balanis, A. S. Spanias, J. M. Capone and T. M. Duman, 'Smart Antenna System Analysis, Integration and Performance for Mobile Ad Hoc Networks (MANETs)', *IEEE Transactions on Antennas and Propagation*, **50**(5), 2002, 571–581.

[27] R. Ramanathan, J. Redi, C. Santivanez, D. Wiggins and S. Polit, 'Ad Hoc Networking with Directional Antennas: A Complete System Solution', IEEE Wireless Communications and Networking Conference (WCNC), 2004, pp. 375–380.

[28] A. Spyropoulos and C. S. Raghavendra, 'Energy Efficient Communications in Ad Hoc Networks Using Directional Antenna', IEEE Annual Conference on Computer Communications (INFOCOM), 2002, pp. 220–228.

[29] T.-S. Yum and K.-W. Hung, 'Design Algorithms for Multihop Packet Radio Networks with Multiple Directional Antenna Stations', *IEEE Transactions on Communications*, **41**(11), 1992, 1716–1724.

[30] A. Nasipuri, S. Ye and R.E. Hiromoto, 'A MAC Protocol for Mobile Ad Hoc Networks Using Directional Antennas', IEEE Wireless Communications and Networking Conference (WCNC), 2000, pp. 1214–1219.

[31] Engim Inc., 'Multiple Channel 802.11 Chipset', Available from <http://www. engim.com/products_en3000.html>

[32] M. McHenry, 'Frequency Agile Spectrum Access Technologies', FCC Workshop on Cognitive Radios, May 2003.

[33] B. Lane, 'Cognitive Radio Technologies in the Commercial Arena', FCC Workshop on Cognitive Radios, May 2003.

[34] J. Mitola III, *'Software Radio Architecture: Object-Oriented Approaches to Wireless System Engineering'*, Wiley Inter-Science, New York, 2000.

[35] J. Kajiya, 'Commodity Software Steerable Antennas for Mesh Networks', Microsoft Mesh Networking Summit, June 2004.

Index